The Complete Dictionary of Television and Film

The Complete Dictionary of Television and Film

LYNNE NAYLOR ENSIGN
and
ROBYN EILEEN KNAPTON

Foreword by Norman Lear

STEIN AND DAY/*Publishers*/New York

First published in 1985
Copyright © 1985 by Lynne Naylor Ensign and Robyn Eileen Knapton
All rights reserved, Stein and Day, Incorporated
Designed by Louis A. Ditizio
Printed in the United States of America
STEIN AND DAY/*Publishers*
Scarborough House
Briarcliff Manor, N.Y. 10510

Library of Congress Cataloging in Publication Data

Ensign, Lynne Naylor.
  The complete dictionary of television and film.

  1. Television—Terminology—Dictionaries. 2. Moving-pictures—Terminology—Dictionaries. 3. Cinematography—Terminology—Dictionaries. I. Knapton, Robyn. II. Title.
PN1992.18.E57   1985   791.43′03   83-42634
ISBN 0-8128-2922-0

ACKNOWLEDGMENTS

This book was begun more than eight years ago, and it is with relief and exhaustion, as well as a hopefully not misplaced sense of pride, that we offer it to the public and to the current and future entertainment industry. Covering everything from words such as "apple box," coined in the early days of silent film, to television jargon of the '70s and '80s such as "backdoor pilot," we present a volume that attempts to compile, define, and standardize the language of television and film. It is the first book in English to do so.

In the course of our research, we very often found that a word that was used in one sense at one studio or television network was used in an entirely different sense elsewhere. Standardization was not easy, but we have done our best not only to determine the most common definition of a word but to discover any other subordinate meanings that are in widespread use. We have included slang since it is as integral to the language of television and film as is the most technical and scientific terminology.

Inevitably in a project of this scope, there will be terms that have been overlooked, and variant meanings that deserved but did not receive consideration. We welcome additions, deletions, and comments of any kind; they can be sent to us care of Stein and Day/*Publishers*.

Both of us wish to acknowledge those who helped us in their special ways. Special thanks...

...from Robyn to:	Amy and Richard Knapton, Barbara Butler, my sister, Rodi, Richard Hatch, and Patrick Picciarelli.
...from Lynne to:	Buzz Ensign, my sons, Joey and Jon Naylor, and Virginia L. Carter.
...from Both of us to:	Sol Stein and Toni Mendez, as well as Carol Haymer, Nancy Grill, Bill Fryer, Daphne Hougham, Sid Spalding, Harry Evans, Embassy Television/Embassy Pictures, 20th Century-Fox, Metromedia, Universal Studios, and Lorraine, without whom we would not have been able to complete this work, and of course... our dear Norman Lear.

FOREWORD

The pages of this book represent a remarkable endeavor. At last, gathered together in one place, we have a first-ever summary of the language of television and film. It is a language which has evolved over the past eighty years to serve an industry that grew at a great pace in a curious kind of isolation. The result has been the evolution of an idiom whose meanings are far from obvious. After forty years in the business, I can still be brought up short on occasion by a bit of industry jargon. I'll be keeping this book by my side! The definitions are simple and to the point and should prove helpful to film and television professionals, to lay people, and to students, alike.

Ours is a fluid industry, by which I mean very few people get their twenty-year pins for meritorious service to one company. On the contrary, people in film and television move from project to project, and from one facet to another of the business with easy regularity. Every move expands one's working vocabulary. Add to this shifting scene the emphasis we have placed over the past decade on bringing newcomers into the field—people with fresh insights and a variety of cultural backgrounds. We need them, yet our specialized language can pose a real barrier to entry and acceptance. The truth is, as I have often suspected, that a newcomer might easily pass for an oldtimer given enough chutzpa (courage, guts) and a thorough, up-to-date knowledge of our working language.

This is a book that saw its first service while still in its pre-galley stage. One of our top people here at Embassy put a network executive on hold while frantically demanding of his secretary the meaning of the expression "long form." As luck would have it, one of our authors was passing by with this handy-dandy reference work in hand. It was the first opportunity for this book to be of service. There will be many more.

Practically everyone with a television set, or who enjoys watching films, will find this text fascinating. It will help with an understanding of many of those words which have begun to cross over into everyday language, as well as serve as an introduction to many facets of the film and television industry.

I am proud to congratulate Robyn and Lynne, whom I've known for many years, on their fine work.

Norman Lear

The Complete Dictionary of Television and Film

A AND B ROLLS
(Also Checkerboarding)

A checkerboard technique whereby alternate scenes are intercut with black leader in such a way that on Roll A scenes are evident, where, on Roll B, black leader is presented. This technique allows multiple exposures for use in superimposing titles, the printing of dissolves from one scene to another, and the achieving of invisible splices.

A AND B PRINTING

Printing from original film, conformed into two rolls, where each scene is staggered and overlapped with black leader, checkerboard style.

A AND B SUPERED TITLES

Titles superimposed over action by double-printing the A roll over the B roll or vice versa.

A LEFT-HANDED CLOCK IS FAR BETTER

A mnemonic sentence used to help indicate whether or not a roll of film has a B-wind position. While holding the roll in the left hand, the film will come off clockwise, with the perforations on the side away from the holder. (See also A-Wind, B-Wind)

A-ROLL ORIGINAL

Original film assembled into a single roll for printing.

A-WIND

A term used in describing the position of the emulsion and perforations of single perforation film. That is, when the dull side is toward the person holding the film, with the tail of the film hanging down from the right side, the emulsion is facing in, and the perforations will be on the edge nearest the person holding the film. (See also B-Wind, Printing Wind)

ABC

See American Broadcasting Companies, Inc.

ABERRATION

Image distortion on film caused by an optical element such as a lens, mirror, or prism.

ABOVE-THE-LINE COSTS
(Also Front End)

Production costs incurred by acquisition of the story rights and by fees paid to the writer, producer, director, and principal cast members.

ABRASION MARKS

Visual defects appearing on film caused by scratches from dirt, emulsion pileups, or grit.

ABSORPTION FILTER

A light filter that allows certain wavelengths of light to pass, while blocking others. It can be used to create mood, special effects, or subject clarity and distinction. (See also Filter)

ACADEMY APERTURE

The size of the frame mask in 35mm projectors and cameras as established by the Academy of Motion Picture Arts and Sciences.

ACADEMY LEADER

A standard film leader containing specific cue marks that aid in projection. The leader, designed by the Academy of Motion Picture Arts and Sciences, is spliced to the beginning of each reel of film. (See also Leader, Projection Leader)

ACADEMY OF MOTION PICTURE ARTS AND SCIENCES

Established in 1927, the Academy of Motion Picture Arts and Sciences is a professional, honorary organization composed of more than 4,500 motion picture craftsmen and artists. Its purposes are to advance the arts and sciences of motion pictures; foster cooperation among creative leaders for cultural, educational, and technological progress; recognize outstanding achievements (see Oscar); provide a common forum and meeting ground for various branches and crafts; represent the viewpoints of actual creators of motion pictures; and foster educational activities between the community and the public at large. The Academy's headquarters are located at 4605 Lankershim Blvd., Suite 800, No. Hollywood, CA 91602.

ACADEMY PLAYERS DIRECTORY
(Also The Directory)

Published three times annually by the Academy of Motion Picture Arts and Sciences, these large volumes contain (for a fee) the actor's picture, agent, manager, or contact. The volumes are divided into several categories, including leading men, leading women, ingenues, character actors, and children. dren.

ACADEMY STANDARDS

Standards of film leaders, camera apertures, projectors, etc., as set by the Academy of Motion Picture Arts and Sciences.

ACCELERATED MOTION

See Fast Motion.

ACCEPTANCE ANGLE

The angle that has enough light to be covered by the camera for a specific shot.

ACE

(1) A spotlight utilizing a 1,000-watt lamp; (2) the abbreviation used for the Association of Cinema Editors.

ACETATE (Also Acetate Base, Safety Base)

(1) A film base consisting of slow-burning cellulose triacetate used to support film emulsion and other coatings on motion picture film; (2) the processed sheet form used as "cels" for animation work.

ACHROMATIC

Any optical equipment, such as a

lens, that has been corrected to prevent chromatic aberration.

ACOUSTICS
(1) Having to do with sound; (2) the properties of a room that effect sound. A room (a concert hall, for example) is sometimes constructed to enhance the acoustics.

ACOUSTICAL FEEDBACK
An audio disturbance, usually a high-pitched screech, that occurs when a live microphone is placed too close to the loudspeaker from which the mike is being amplified.

ACT
(1) To perform a part in a production; to take on the characteristics specified in written dialogue and appear to become, or give life to, that which is written: to act the part of an old man, or other distinct character; to act angry, in love, hysterical, or other emotional elements; to act in a certain style, such as the Stanislavski method: (2) a rehearsed, packaged performance, complete as is: a magician's act, a clown act, a singing act; (3) one of the major divisions or sections of a television script, comprised of scenes.

ACT BREAK
A definite break or shift in the action or direction of the plot; in television, commercials are usually scheduled at an act break.

ACTING
The physical interpretation of a script's dialogue by a performer, using speech, movement, and gestures.

ACTION
(1) The physical movement or business of the subject(s) within the camera's field of view; (2) "Action" —the command given by the director to begin acting.

ACTION CUTTING
The editing of a film in such a way that the action on the screen appears continuous, making changes in camera position or setup undetectable. This effect is commonly achieved by having the action of the second shot begin with the end action of the first shot, overlapping in such a way so that it is not apparent to the viewer that the camera was stopped. Two cameras can achieve the same effect by switching from one to the other at the proper cue.

ACTION FIELD
(Also Field of Action)
The area recorded by the camera upon which the action to be photographed takes place.

ACTION PROPERTIES
(Also Practical Props)
Props that are actually used by the actors during a scene, such as a telephone used to make a call, a dish that is thrown, or a gun that is fired, as opposed to stationary props used only to dress the set (flowers, bric-a-brac, etc.). (See also Props)

ACTOR
An individual who interprets and

performs a role or part in a production.

ACTOR'S EQUITY ASSOCIATION
A guild whose jurisdiction governs live stage productions.

ACTUANCE
The capability of a lens or film to give clearcut edges to the subject being filmed.

ADAPT (Also Adaptation)
(1) To rewrite a story from one medium to another, such as from a book to a screenplay; (2) to attach one component to another with which it is normally incompatible by means of an adapter.

ADAPTER
An attachment that allows one piece of equipment to be connected with another piece of equipment or power source.

ADAPTATION
See Adapt.

ADDED SCENES
Scenes or shots added to a script after principal photography has been completed.

ADDITIONAL DIALOGUE
Lines of dialogue added to a finished script just before, during, or after filming.

ADDITIVE PRINTER
(Also Additive Color Printer)
A printer that uses three colors of light (red, green, and blue) to print from color originals, or intermediates, to produce a custom composite color; each color is controlled separately to create the desired effect. This is done at the time of exposure on the print stock.

ADDITIVE PROCESS
(Also Additive Color Process)
Although not often used today, this process offers a way of achieving natural color through the use of light filters and black-and-white film. First, a separate black-and-white negative is processed with each of the three primary color light filters. A black-and-white positive transparency is made from each negative. The three positive transparencies are then projected simultaneously, each through its primary color filter. The resulting screen image is in color. (See also Subtractive Process)

AD-LIB
See Improvise.

AERIAL CINEMATOGRAPHY
Filming done with cameras mounted on airplanes, helicopters, or other airborne vehicles, or hand-held by an operator inside the aircraft.

AERIAL IMAGE ANIMATION STAND
A specially constructed stand from which an animation projection system can produce a mid-air or aerial image, thus allowing cell animation to be filmed over live action when the animated image is projected amidst the action.

AERIAL MOUNT
Any one of the numerous supports designed for use in aircraft for holding cameras. (See also Copter Mount)

AERIAL SHOT
A shot filmed from any aircraft or from any means of free suspension in the air. (See also Shot)

AFFILIATE
One of the independently owned television stations contracting with one of the three major networks (ABC, CBS, or NBC) to show that network's programming during certain time slots. Each network has approximately two hundred affiliates. (See also Network)

AFI
See American Film Institute.

AGENT
A person and/or company who represents a client for the purpose of soliciting work for the client. The agent also negotiates contracts between the client and those desiring his/her services. Today, agents work for writers, directors, and producers, as well as performers.

AGITATION
Turbulence deliberately caused during film processing baths to assure an even distribution of chemicals.

AIR AGITATION
The use of compressed air to create turbulence in film processing baths.

AIR DATE
The date on which a show will be broadcast to the general public.

AIR GAP
The tiny space between the playback heads of a tape recorder.

AIR PRINT (Also Air Master)
The final approved print of a tape or film ready to be aired.

AIR SHOW
The second, or final, show taped or filmed before a live audience (the first taping/filming is the Dress Show); the air print is sometimes comprised of scenes from the dress show and the air show; (2) a show as aired.

AIR SQUEEGEE
A device used during the developing process to remove water from film by blowing continuous streams of air onto the film as the film goes from the final wash to the dryer.

ALLEFEX MACHINE
Trade name for equipment used during the days of silent film to produce the sound effects while the film was being shown; the system became known by its trade name.

ALLIGATOR CLAMP
See Gator Grips

AMBIENT LIGHT
Soft, diffused, or scattered light surrounding a subject, created by deliberately lighting reflecting surfaces. Sometimes called the "over-

39" light, as it softens facial features of older performers.

AMERICAN BROADCASTING COMPANIES, INC.
ABC, one of the three major U.S. television networks, which has over two hundred affiliated stations.

AMERICAN FILM INSTITUTE (AFI)
Established in Los Angeles in 1968 by the National Endowment for the Arts as an independent, nonprofit organization to preserve and advance the art of television and film. The AFI gives assistance to new American filmmakers, provides guidance for educators, preserves films, operates a publishing arm for film books, periodicals, and reference works. The headquarters address is 2021 No. Western Avenue, Los Angeles, CA 90027.

AMERICAN NATIONAL STANDARDS INSTITUTE
(Also ANSI, American Standards, ANSI Standards, Standards)
A nonprofit organization established to standardize international specifications in the motion picture industry. Size, shape, volume, and definition standards are set for such things as film, frame placement, and leaders.

AMERICAN RESEARCH BUREAU
See ARB.

AMERICAN STANDARDS
See American National Standards Institute.

AMPAS
See Academy of Motion Picture Arts and Sciences.

AMERICAN SOCIETY OF CINEMATOGRAPHERS (ASC)
An honorary association of motion picture camera people, headquartered in Hollywood.

AMPERAGE (Also Amps)
The amount of current that passes through an electrical circuit, equivalent to wattage divided by voltage.

AMERICAN STANDARDS
See Standards.

AMPERSAND (&)
When appearing between two or more names in a credit, it means the two (or more) names comprise one writing unit or team and split the fee. If "and" appears between the names, each receives a separate unit fee.

ANAMORPHIC IMAGE
An image compressed (usually horizontally) by an anamorphic lens.

ANAMORPHIC LENS
A lens designed to compress horizontally an image twice as wide as it is high within a standard frame. A correcting lens is used during projection to "unsqueeze" the image, thus producing a wide-screen effect.
(See also Lens, Spherical Optics, Unsqueeze)

ANAMORPHIC RELEASE PRINT
A release print of a film whose

image has been compressed horizontally.

ANASTIGMAT LENS
An optically corrected lens for preventing astigmatism. (See also Lens)

ANGLE, CAMERA
See Camera Angle.

ANGLE OF APPROACH
(1) The point from which the director shoots the action. Also called Angel of Vision; (2) the angle from which the audience views the film.

ANGLE OF VIEW
The total vertical and horizontal spectrum covered in the field of action by a camera.

ANGLE OF VISION
The camera angle at which a shot is photographed.

ANGLE SHOT
A shot taken from a different angle, continuing the action from the preceding shot. (See also Shot)

ANIMATION
Drawings that are photographed frame by frame (see also Cel), giving the illusion of movement. The most recognizable form of animation is the cartoon. Animation is also used in live-action films as special effects, or to enhance special effects, e.g., explosions, weaponry fire, monsters, cloud formations, etc. Animation was used in such live-action films as *Star Wars*, *Clash of the Titans*, and *E.T.* (See also Barsheets, Cartoon, Cel, Computer, Cutout, Cycle, Dissolve, East, Hold Cel, In-betweener, In-betweening, Inker, Inking, Limited, North, Ones, Opaquer, Peg Board, Pencil Test, Pin Screen, Pixilation, Platen, Punch, Puppet, Registration Pegs, Scratch-Off, Silhouette, South, West.)

ANIMATION BOARD
Any of the numerous tools used in preparing animation artwork, including a flat base with the standard pegs that hold the animation cels.

ANIMATION CAMERA
A camera capable of shooting single frame exposures, either adapted or specifically designed for the purpose of shooting animation. (See also Twos)

ANIMATION CAMERA OPERATOR
A photographer specializing in the use of an animation camera and its related equipment.

ANIMATION STAND
A device that supports a camera being used to film animation. The stand has the ability to raise or lower a camera in the precise steps necessary to shoot an artwork peg board or platen containing the cels to be photographed.

ANIMATOR
The individual whose responsibility it is to determine the amount of change needed in drawings used for animation.

ANSWER PRINT
(Also First Trial Print)
The first positive print of a finished film—that is, a completely

edited, dubbed, and scored film, generally used by the producers/director to determine what, if any, changes are needed prior to release.

ANTENNA
Equipment created to transmit or receive broadcast signals.

ANTHOLOGY FILM
A full-length feature film comprised of excerpts from several films, or a connected group of complete short films.

ANTIABRASION COATING
A coating, applied by the manufacturer to the base side of film stock to reduce the risk of scratching or damaging the film during handling.

ANTIGROUND NOISE
An electronic system that reduces area or surface noise in optical sound when there is little or no sound connected with the action.

ANTIHALATION BACKING
(Also Antihalation Coating, Rem-Jet Backing)
A coating, applied by the manufacturer to the base side of film stock, which absorbs light that has leaked, thereby reducing the amount of light reflected onto the emulsion.

ANTIHALATION DYE
A gray dye added to the film base to reduce light leakage.

ANTIHALATION UNDERCOAT
Antihalation material layered between the emulsion and base of certain films.

APERTURE
The opening (or iris) on a lens through which regulated amounts of light are admitted. The four main types are: (1) lens aperture—the iris opening that allows light to pass through the lens for controlled exposure of the film; (2) camera aperture—the mask opening that delineates the area of each frame; (3) projector aperture—the mask opening that delineates each frame projected; (4) printer aperture—the controlled opening through which light passes to expose film being reproduced. (See also Iris)

APERTURE PLATE
A metal plate mounted between the lens and the film in a camera or projector that regulates, by cut-outs in the front of the plate, the shape of the image exposed onto the film (in a camera) or defines the frame of film to be projected.

APPLE, BAKER, DOUBLE APPLE, ETC.
Words used to replace letters for clarity when slate numbers are spoken into the audio track. Apple refers to the letter "a," baker to "b," and so on.

APPLE BOXES (Also Riser)
Boxes used to raise actors, props, furniture, etc., into a better position during a shot. Usually wooden and in three sizes: full, half, and quarter. Apple boxes are so named because, in the early days of filmmaking, someone would grab a real apple

box for a short actor to stand on so he would appear taller in a scene. The term stuck. (See also Trench)

APOCHROMATIC LENS
A lens that has been corrected by the manufacturer to prevent spherical and chromatic aberration. (See also Lens)

APPROACH
An order from the director for the camera to move nearer to the subject.

ARB (Arbitron)
One of the major ratings/research companies that surveys an audience's viewing preferences for television stations on a local or national level. ARB also conducts surveys for radio stations.

ARC
(1) The curved path followed by a camera dolly; (2) the line of story progression; (3) the spark produced when electric current jumps across an air gap.

ARC OUT
(1) An order to a dolly pusher to move the dolly on a curve away from the action; (2) an order telling an actor to cross in front of the camera on a curved path.

ARC LIGHT
Large, powerful lights used either on location or in the studio to augment or simulate sunlight.

ARCHETYPE
A story or film with an identifiable structure, symbolism, or pattern similar to other works in its genre.

ART DEPARTMENT
The center or group of people responsible for all graphic material used during filming.

ART DIRECTOR
The person who arranges for and supervises set design and preparation, after assessing the staging requirements of a production.

ART FILM
(1) An avant-garde or experimental film that uses bizarre or new techniques of production, plot or performance; (2) an informational film about or revolving around one of the major art forms.

ART HOUSE
A commercial theater that specializes in showing art, foreign, cult, or classic films.

ART STILL
A picture painted or prepared by the art department to be used as a slide for rear projection on a set or as a prop, decoration, or for some other staging purpose.

ARTWORK
A term encompassing all areas of graphic art used in connection with a film, including titles, animation, set decoration, and diagrams. (See also Graphics)

ARTIFICIAL LIGHT
Light from any source other than natural light from the sun or sky;

any electric, battery, or fueled light source.

ARRI
The abbreviation commonly used for an Arriflex camera; "Ar" from Arnold and "ri" from Richter, the two inventors.

ARRIFLEX
The most widely used film camera today, the Arri can be used in a fixed position or as a hand-held camera. The Arri has replaced the Mitchell in popularity.

ASA SPEED (Also ASA Exposure Index, ASA Rating)
A term indicating the degree of sensitivity to light as established by the American National Standards Institute, Inc.

AS-BROADCAST SCRIPT
The verbatim dialogue of a television script as it has been recorded and as it will air.

ASC
The abbreviation used for the American Society of Cinematographers.

ASPECT RATIO
(Also Screen Ratio)
The ratio of frame width to frame height as regulated by changing the cutout shape in the aperture plate or by using an anamorphic lens. The standard dimensions are based on a 4 to 3 relationship: for 8mm, super 8mm, regular 16mm, and 35mm frames, this aspect is 1.33 to 1 (usually expressed as 1 x 1:33 or 1.33:1). The aspect ratio for Cinemascope is 2.35 to 1.

ASSEMBLE
(Also Assembly, Stringout)
The first stage of film editing in which all shots are connected in order as written in the script. (See also Workprint)

ASSISTANT ANIMATOR
See In-Betweener.

ASSISTANT CAMERAMAN
Directly responsible to the camera operator for the set-up and maintenance of the camera and its associated equipment, including lens stop and focus for all takes.

ASSISTANT DIRECTOR
(Also Associate Director)
In film, the individual who handles production problems. The A.D. does not concern himself specifically with the art of directing. He performs tasks in connection with the production that can be accomplished without the supervision of the director. These include scheduling, supervising crews, conducting run-throughs, and staying near the director during shooting in order to assist with such problems. In television production, the individual who functions in this capacity is called the associate director.

ASSOCIATE DIRECTOR
See Assistant Director

ASSOCIATE PRODUCER
(1) In television production, the associate producer has overall control and supervisorial responsibility of all aspects of production, including hiring of support personnel, and

assists the director with editing during sweetening sessions; (2) in film production for television and feature films, the associate producer primarily supervises the postproduction processes.

ASSOCIATIONAL EDITING
See Relational Editing.

ASTIGMATISM
An optical defect, which can be corrected by the manufacturer, of a lens in which light rays passing through the lens fail to converge properly, causing unclear focus.

ASYNCHRONOUS SOUND
Sound connected to, but not synchronized with, the action being filmed. Sometimes the source of the sound is not visible but is assumed to be there. Examples are the sound of crickets added to a scene around a campfire, the mooing of an off-camera cow in a barn, or the sound of dry leaves being crushed by footsteps.

ATMOSPHERE
The total combination of characteristics that set the tone for the action. Atmosphere can be communicated by lighting effects, costumes, and sets as well as by the mood or emotional tone set by the actors.

ATOMIC BOMB WIPE
See Wipe.

ATTRACTION
See Hard Ticket Roadshow.

AUDIENCE RATING
An evaluation or scale of judgment used by those viewing a production for the purpose of rating.

AUDIENCE SHARE
See Share.

AUDIO
(1) The sound portion of a production; (2) any part or all of the equipment and facilities related to sound and its reproduction.

AUDIO FILTER
See High-pass Filter.

AUDIO TAPE (Also Magnetic Tape, Quarter-inch Tape)
Record tape with a coating capable of absorbing and playing back magnetically recorded sound. Audio tape is readily available in cassettes or on reels, and the sizes include 1/8-inch, 1/4-inch, 1/2-inch, 1-inch, and 2-inch widths.

AUDITORY FLASHBACK
A flashback that uses sound only to recall a previous event, or passage of dialogue. The visuals remain in the present, while the auditory flashback is occurring.

AUDITORY PERSPECTIVE
The discernment of direction, distance of sound, as estimated or measured by the ear.

AUDITION
A try-out or test performance by anyone desiring a job in theater, TV, music, or films. (See also Camera

AUTOMATIC EXPOSURE

Test, Personality Test, Screen Test, Sides, Test)

AUTOMATIC EXPOSURE
A camera device that automatically adjusts the lens iris for brightness.

AUTOMATIC GAIN CONTROL
An automatic electronic device that can hold a variable sound such as applause, screams, laughter, or drums, at a certain volume or output level.

AUTOMATIC SLATE
(Also Automatic Slating)
See Electronic Clapper.

AVAILABLE LIGHT
Existing light not supplemented by use of studio lights.

AVAILABLE LIGHT FILMING
Filming done with existing light sources with no supplemental light added.

AVANT-GARDE FILM
A film exhibiting new, advanced ideas or techniques, or an unconventional style.

AWARDS
See Emmy, Oscar.

AXIS
An imaginary line drawn through the action being filmed. Crossing the line from one shot to another results in the action appearing to reverse its direction on the screen. (See also Crossing the Line, Imaginary Line)

B

B PICTURE
A colloquial expression for a low-budget feature film, or one made with inferior production values in writing, casting, location selection, or camera techniques.

B-WIND
A term used to describe the position of the emulsion and perforations of single-perforation film. Film is either A-wind or B-wind; B-wind indicating that the position of the emulsion and perforations is opposite, or reversed, from A-wind. That is, when the film is held vertically, the end of the film comes off the reel downward from the right side, with the perforations on the edge away from the person holding the roll and with the base side of the film facing up.

BABY
A small, focusable Fresnel lens spotlight that uses a 500, 750, or 1,000-watt bulb.

BABY LEGS (Also Baby Tripod)
A very short tripod used for holding a camera while shooting low angle shots or table top shots.

BABY PLATE
(Also Baby Wall Plate)
A small mount for a baby spotlight.

BACK, TO
To bring in background music or sound effects behind a scene.

BACKLIGHT
A lighting technique in which the principal light source is behind the subject, creating a soft halo effect or silhouette. Sometimes referred to as Rembrandt lighting. In the studio, backlighting is usually created by a spotlight positioned behind and above the subject. In outdoor lighting situations, the sun is often used as a backlight.

BACK LOT
Generally, an area located away from the studios and office buildings, however, still on the lot; used for shooting exteriors. Ordinarily various "streets" are erected on a back lot, ranging from New York, circa 1935, to Tombstone, Arizona, 1873; these streets can be changed to suit many film location needs. (See also Studio Lot)

BACK PROJECTION
See Rear Projection.

BACKDOOR PILOT
A movie made for television that can be considered a pilot episode for a series if a network or buyer is interested in buying a series based

on the characters introduced in the MOW.

BACKDROP (Also Drop)
A painted or photographed background used to simulate a natural landscape when filmed. Backdrops are curtains, cycloramas, or flats, and are used behind windows and doors of sets to create, for example, a cityscape outside the set of an apartment. Some backdrops are very elaborate with, for example, twinkling lights to simulate city lights.

BACKGROUND
(1) That which is seen behind the action; (2) music or sound effects played at a lower volume than the rest of the action.

BACKGROUND ACTION
Action that takes place behind the main action occurring in the foreground, sometimes directed by an assistant director.

BACKGROUND LIGHTING
Lighting created to highlight people, objects, or areas in the background of a scene to create a mood or reinforce the feeling on a set; also, artistic lighting patterns in the background of a shot.

BACKGROUND MUSIC
The music score or soundtrack heard in the background of a feature film or television production. In major theatrical films or television productions, the music is carefully written, arranged, or selected to fit and reinforce the action or the emotional tone. Background music in nontheatrical films, such as an educational film, is usually not scored to match the action, but is usually selected from a prerecorded source such as a record or tape. (See also Organ Background Music)

BACKGROUND PLATE
A glass slide with a photograph attached, used as a background in a rear projection system.

BACKGROUND SOUND
The sound in the background of a shot that creates a sense of reality, such as normal room noises (see also Room Tone) or crowd sounds and appropriate sound effects. It is either created naturally on location or added during the audio mix. (See also Mix)

BACKING
Any coating, such as a rem-jet antihalation coating, applied to the back of film stock. (See also Base)

BACKING REMOVAL
The step to remove black antihalation backing from film. It is first chemically softened and then buffed off before complete processing.

BACKUP INTERLOCK SYSTEM
Magnetic soundtracks used with a pick-up recorder that can be erased easily for rerecording during a mix.

BACKUP SCHEDULE
A schedule listing shots to be filmed that can be substituted for, and shot in place of, the original

schedule if changes must be made because of unforeseen technical problems, bad weather, or cast illness.

BAFFLE
(1) A portable acoustic wall or batting situated near an area of sound recording on a set or in a studio, in order to control reverberation during recording; (2) a plate, screen, or enclosure for a loudspeaker unit designed to regulate the movement of sound.

BALANCE
(1) The degree of brightness compared to shadow in a picture when referring to light; (2) the arrangement of microphones, instruments or sound sources to obtain the best balance for recording when referring to sound or music. (See also Equalize)

BALANCE STRIPE
A narrow band of magnetic coating placed on the edge of magnetically striped film stock opposite the magnetic sound track stripe. It creates a balance for the sound stripe and enables the film to lie flat when passing over magnetic heads. This balance stripe can also be used for additional audio or cueing information.

BALANCED PRINT
A print that has been color corrected or graded. (See also Grading)

BALLAST
A unit that controls the flow of electric current to another source, such as a lamp.

BALOP
A colloquial term for a larger-than-normal slide; the term comes from a trade name for a slide projector, Baloptican.

BALOPTICAN
A trade name for a slide projector that uses large-size slides. (See also Balop)

BANDWIDTH
Measured in hertz (Hz) or cycles per second, bandwidth is the difference between the highest and lowest frequency components that can be carried by a telecommunications system. For example, telephone voice transmission requires a bandwidth of approximately 3,000 cycles per second (3 kHz), television channels need six million cycles per second (6 mHz). Modern cable systems usually use from 50 to 300 mHz.

BANK
(1) To arrange lights in a group; (2) the group of lights itself, or light bank.

BAR SHEETS
(Also Lead Sheets)
Used primarily in animation, a chart indicating the dialogue spoken and the duration and corresponding number of film frames used for each syllable and pause. (See also Animation)

BARNDOORS (Also Flippers)
Metal shields, either two-leaf or four-leaf; hinged to the front of a lighting instrument in order to shape the light beam, restrict the

amount of light emitted, create patterns, or prevent microphone shadows.

BARNEY
A padded exterior camera cover, usually like a blanket and sometimes with heating coils, used on the camera to reduce camera noise; not as compact or fitted as a blimp.

BARREL DISTORTION
An image distortion in which the sides of square objects appear through a lens to be bent outward.

BARREL MOUNT
A lens mount complete with all necessary lens elements, usually detachable and extending outward from the camera.

BARREL SHUTTER
Used on some projectors, a cylindrical rotation shutter that has two openings opposite each other.

BASE (Also Backing, Support)
A thin, flexible material, most commonly cellulose triacetate, that acts as a base for coatings of light-sensitive film emulsion, magnetic recording compound (iron oxide), or other coatings. (See also Acetate, Emulsion)

BASE DENSITY
The density of the unexposed area of negative film after fixation.

BASE DOWN, BASE UP
(1) Terms that indicate which direction the film base faces for proper threading position for that particular film; (2) to describe the normal position of certain lamps.

BASE LIGHT (Also Base Illumination, Set Light, Foundation Light)
The first lighting done on a set. Usually a soft light emanating from large diffused floodlights, which raise the general level of light, before the second set of lights, the key or modeling lights, are positioned.

BASE PLATE
A pedestal mount with an upright shaft upon which a light can be affixed.

BASE SIDE
The under side of film, opposite the emulsion side.

BASE-TO-BASE SPLICE
A film splice in which the base side of one film is overlapped on the base side of the film to which it is to be spliced; used, for example, in checkerboard editing to eliminate the need to scrape emulsion.

BASE-TO-EMULSION
A term that indicates a change in the standard winding of film; a base-to-emulsion print from an original would have the emulsion and soundtrack in B-wind position. (See also B-Wind)

BASIC SET
A set not furnished with props.

BASS BOOST
An electronic component used to

increase the intensity of low frequency sound.

BASS ROLLOFF
A gradual attenuation or reduction in the bass, or low frequency, levels of a sound or music track.

BASTARD AMBER
A much-used red-orange gelatin; Brigham's No. 62, light scarlet. (See also Gel)

BATCH NUMBER
(Also Emulsion Number)
An individual identification number assigned to each batch of all raw stock manufactured at the same time; each batch possesses identical sensitivity and color characteristics.

BATH
Any solution or rinse used to bathe film or prints during developing or printing stages.

BATTEN
A pole suspended horizontally across the field of action from which lights and scenery can be hung.

BATTERY BELT
A belt, which can be worn by camera operators, with batteries and connections for camera power cables. When worn, it leaves an operator's arms free for other tasks.

BAYONET MOUNT
A camera lens mount that holds the lens in place with a snap-lock device, as opposed to a threaded, screw-on type mount.

BEADED SCREEN
A screen coated with plastic or glass beads that possess a high reflectance capability.

BEAM SPLITTER
A prism or partial mirror capable of reflecting part of a light beam and passing through the rest. A beam splitter is used to separate colors and to produce two images in two separate places.

BEAT
(1) Pause or hesitation between words or action, "one beat" would be a very quick break in the action, "two beats" would be slightly longer; (2) story accents in the progression of a script.

BEEP
A tone, usually 1,000 Hz, aligned with a visual point of reference, in order to synchronize the soundtrack exactly with the picture during editing, mixing, and printing.

BEHIND-THE-LENS FILTER
A gelatin filter secured by a metal frame that is inserted in a slot between the back of the lens and the film.

BELLY BOARD
A flat board upon which a camera can be mounted for low angle shots.

BELOW-THE-LINE COSTS
All costs relating to actual production. For example, production

staff, camera, sound recording, art direction, set construction, set operations, electrical, wardrobe, set dressing, props, makeup and hair dressing, film processing, special effects, locations, transportation, studio rental, retakes, and added footage. (See also Above-the-Line Costs)

BEST BOY
First assistant to the head gaffer.

BETAMAX
A video recorder/playback unit manufactured by the Sony Corporation that uses 1/2-inch magnetic videotape in a cassette smaller than a VHS tape. (Betamax, along with VHS, is used mostly in households for home entertainment. Three-quarter inch tape systems are used most commonly in professional production situations.) (See also VHS)

B.G.
Abbreviation for background.

BI-DIRECTIONAL
A term used to describe a microphone or a microphone pickup pattern in which a mike picks up sounds emanating from the front and back but not from the sides.

BIAS
A method of reducing noise and distortion by applying a high frequency alternating current to magnetic recording heads that moves the audio signal to the linear portion of the magnetic recording curve.

BIBLE
The complete information regarding a script's characters, story lines, locations, and details that can be referred to as a script is written. Used especially in episodic television, a bible can be used for any type of production. The bible contains details that help to keep a script consistent in details. A character's likes, dislikes, and background are examples of the type of information found in a production bible. When writing a character's realistic reaction to a scene about, for example, the fact that a minority is moving into the neighborhood, the bible might be checked to see if the character has reason to show prejudice. Also, a character wouldn't serve ham at a dinner if the bible were checked and showed that the person was raised an Orthodox Jew. This can be imperative in a long-running series. For example, if in one episode, a character has made a big point about hating bologna, it would be inconsistent in episodes down the line to show the character enjoying a bologna sandwich and not to offer an explanation. (See also Story Editor)

BICYCLING
The use of one print for staggered showings in more than one theater. The term originated in the early years of movies, when reels of film were transported by bicycle from one theater to another.

BILATERAL VARIABLE-AREA SOUNDTRACK
(Also Bilateral Track)
A variable area soundtrack in which the modulations run symmetrically lengthwise down the track.

BILLING
See Credits.

BIN
A container used for temporary holding or storage of strips of film during editing. Resembling a large wastebasket or barrel, a bin is lined with soft cloth to prevent scratching, and has a rack attached from which film strips are suspended with their ends dangling into the cloth-lined bin.

BINDER
The gelatin coating that causes the light-sensitive silver particles to adhere to a film emulsion.

BIPACK PRINTING (Also Bipack Contact Matte Printing)
A printing technique of running two pieces of film in contact for matting or double exposure using a contact printer or an optical printer to print two bipack films, sandwiched together, onto a third piece of film or printer stock. Bipacking can be done in a camera or a printer.

BIPACK DOUBLE PRINT TITLES
The use of a bipack system to print titles on a duplicate film print that results in the action and the titles appearing on one print. Depending on whether the film stock is negative or positive, a duplicate can be made with either white titles or black titles superimposed over the action. (See also Bipack Printing, Title)

BIRDSEYE
A colloquial term for R and PAR lamps, from the name of the inventor, Clarence Birdseye. (See also Molepar)

BIT PLAYER
An actor who regularly does very small parts, whether by choice or because they are the best he can find. Many big stars began as bit players until their big break.

BIT PART
A very small role in a film, without the attention given to a "cameo" performance. A cameo, although a small part, is more significant than a bit part. Bit parts are often performed by day players. (See also Cameo, Day Player)

BLACK
A dark screen; black videotape or film between scenes; part of a directive regarding the cutting of a picture, i.e., "fade to black."

BLACK AND WHITE
Without color; photographed images translated into degrees of whites, grays, and blacks, representing the natural colors of the photographed images in a monochromatic array of gray tones.

BLACK COMEDY
A film that treats serious themes such as death or sacred issues in a comedic manner.

BLACK LEADER
(Black Opaque Leader)
A strip of black film leader free of pinholes, used to block printer light when conforming original film for A and B roll printing. (See also Leader)

BLACK LIGHT
See Ultraviolet.

BLACK NET
A method of reducing light with a minimum of diffusion by using a black screen or black netting.

BLACK TRACK PRINT
Image without sound; a print containing video only.

BLACKLIST
See Hollywood Blacklist.

BLACKOUT
(1) A quick cut off of all lights or picture (See also Black); (2) a short skit with an unexpected twist or ending.

BLACKOUT SWITCH
(Also Master Switch)

A master switch controlling all the lights on a set.

BLACKSPLOITATION FILM
(Also Blaxploitation Film)

An exploitation film using black performers in stereotypical roles, often with thin, sensationalistic, crime-oriented plots. (See also Exploitation Film, Sexploitation Film)

BLEACH (Also Bleach Bath, Bleach Solution)
A step in the reversal process in which a chemical solution is used to dissolve and remove metallic silver from a film's emulsion without disturbing the remaining undeveloped silver salts.

BLEACHING
The conversion of the metallic-silver image into halides (later removed during fixing), a necessary step in processing color film.

BLEEDING
(1) To print a picture so that the image is printed over the edges of a usual border or boundary; (2) a visual defect that occurs in the developing process on the edges of high-density and low-density images, when the definition between the images is weakened by harsh development, especially noticeable around images such as a silhouette against a light sky.

BLEND-LINE
(Also Matte-Line)

The separating line between a matte image and the image of the action.

BLIMP
A somewhat fitted casing used to surround a camera and soundproof it, to prevent unwanted mechanical camera noise from being recorded; more fitted than a barney and sometimes built-in. (See also Zoom Blimp)

BLIMPING
The soundproofing material used in a blimp.

BLIP TONE
A quick popping sound that occurs on tape or optical soundtrack; a bloop or sync pop.

BLOCKBUSTER
(1) A tremendously popular film, whether made with a large or small budget, that breaks box-office records or has huge audience appeal; (2) a big-budget film spectacular released initially in a limited num-

ber of theaters that will charge higher than normal admission prices. (See also Sleeper)

BLOCKING

The planned movement or positions of the performers and cameras to be followed during the entire production to keep the performers in camera range, and achieve the best angles. Often positions are "chalked off" or "marks" are taped to the floor to achieve exactness. Blocking is the second step of rehearsal (following the reading) and is designed by the director.

BLOCKING THE SCENE

To create the plan of movement or positions of the performers and cameras that will be used when a scene is shot.

BLOOP

(1) A quick clicklike noise, or blip, that is the result of a splice on a film with an optical soundtrack passing over the projector lamp, or of a splice on magnetic track passing over the heads of a playback unit; (2) to reduce the sound of a bloop by opaquing the track section of a positive film splice; (3) to punch a specially shaped hole with a blooper in the soundtrack area of a film negative for the same result; (4) to erase an unwanted sound from a magnetic soundtrack carefully by hand with a small magnet. Both of the latter definitions are also called deblooping. (See also Blip Tone)

BLOOPING INK

Any dark, opaque ink used to cover unwanted portions of a soundtrack. By applying blooping ink to a soundtrack, the covered portions will not reproduce.

BLOOPING TAPE

Tape used in editing to cover unwanted parts of a soundtrack, which means the sounds under the covered track will not be duplicated.

BLOW-UP

(1) To make an enlargement of a print, frames, or part of a frame of film on an optical printer, or to duplicate on a larger gauge film stock; (2) a phrase used sometimes when an actor makes a mistake when speaking a line of dialogue.

BLUE COMETING

Caused by metallic contamination during the processing baths, small light-bluish spots that appear in the developed color emulsion.

BLUE MOVIE (Also Blue Film)

A slang term for a pornographic film.

BLUE SCREEN PROCESS

The technique of photographing action in front of a blue background in order to make a matte.

BLUR PAN

See Swish.

BNC

The standard sound camera manufactured by the Mitchell Camera Company.

B.O.

Abbreviation for box office.

BOARDS

(Also Production Boards)
One of the earliest preproduction

tasks, the breakdown, scene by scene of a script, identifying transportation and vehicles needed; special effects; principal cast members (indexed by number, and listed in the Production Book); extras; number of interiors and exteriors to be shot; the number of pages of script to be shot in a day. This information is then transferred to strips of paper, one for each day of shooting. From this board information is available in order to rearrange the shooting schedule if necessary, and for use in drawing up various necessary production schedules such as the daily shooting schedule, or the day out of days. (See also Production Book)

BOARD BREAKDOWN
See Production Book.

BODY BRACE
A camera brace, which attaches to a camera operator's shoulder and waist, that allows the operator's hands to remain free to follow focus and zoom.

BOFF
See Plot Gimmick.

BOMB
A colloquial term for a film that is a box office and audience failure.

BOOK
(1) A script of a production, especially the dialogue or story of a musical; also, a copy of a script containing the director's and stage manager's blocking diagrams, notes, etc.; (2) two flats hinged together that can be folded easily for mobility.

BOOM
(1) A sound dolly with an adjustable pole holding a microphone that can be moved silently around the set and extended over the performers' heads; (2) a sturdy cranelike support, upon which a camera is mounted, that provides complete vertical, horizontal, diagonal, pivotal, and lateral movement for the camera and operator during filming.

BOOM OPERATOR
The crew member who is responsible for the operation of a microphone boom.

BOOM SHOT (Also Crane Shot)
A shot by a camera mounted on a boom or crane that allows total mobility of the camera.

BOOM UP, BOOM DOWN
A command given to the boom operator to move the camera or microphone boom in a vertical direction.

BOOMERANG
A device that holds gels on the front of lights.

BOOMY (Also Tubby)
A colloquial description of reproduced sound that either lacks definition or has too much bass tone (low frequency).

BOOSTER LIGHT
An arc lamp used on locations for

enhancing the daylight needed to get a shot; any auxiliary artificial light used to augment daylight photography, when the sun is used as the key light. It is frequently used to improve shadow detail.

BOOT
A cover made of leather or fabric used to slip over a tripod head for protection when the tripod is being transported or stored.

BOOTH
Short for control booth. (See also Control Booth)

BOOTLEG
The illegal reproducing of a film or production for commercial sale; also a copy of the reproduction. (See also Pirated Print)

BORDER LIGHT
See Striplight.

BOTTOM PEGS
The pegs on an animation board that are nearest an animation crane operator who is facing the equipment.

BOUNCE LIGHTING
Light from conventional lighting equipment aimed at the ceiling and walls to achieve very diffuse lighting.

BOX LUNCH
Lunch in a box, handed out on location shoots to the cast, crew, and production personnel.

BOX OFFICE
(1) Income from a film's distribution, rentals, and sales, which is an indication of a film's overall popularity and success; (2) a small area in front of a theater for the purpose of selling and distributing tickets to a production.

BOX-OFFICE DRAW
A term describing the potential of a film or star to draw an audience into theaters. (See also Box Office, Star System)

BRACKETING
The technique of exposing film not only at the f-stop indicated by the light meter, but also at several other greater and smaller f-stops to be certain of attaining a shot at a perfect exposure.

BREAK
(1) A colloquial term for a chance; an opportunity to do a part, usually small to start. Sometimes an actor's first break comes by getting a bit part in a major feature film. The term applies industry-wide for all major positions, but especially, actors, also directors and producers. (2) released, as when a news story breaks to the media.

BREAKAWAY
A prop, structure, or piece of furniture specially constructed so that it collapses, or breaks, in a realistic manner on cue. Any prop that can be broken without injuring performers. Also used as an adjective as in breakaway glass or breakaway chairs.

BREAKAWAY GLASS
Thin plastic molded into the

shape of a particular prop, such as a bottle, or structure, such as a window. The special effects department creates such articles so that they can be used to simulate glass breaking without danger to the performers. The sound effect of the breaking glass is added later in editing. (See also Breakaway)

BREAK IT UP (Also Coverage)

Filming the master scene from other angles, generally close-ups.

BREAK-UP

Audio or video interference, such as static.

BREAKDOWN
(Also Breaking Down)

(1) A director's analysis, shot by shot, of action to be photographed; (2) in editing, the planned separation, omissions, and rearrangement of shots before actual cutting; (3) to tear down a set on a sound stage or location. (See also Set up)

BREATHING (Also In-and-Out-of-Focus Effect)

What appears to be the moving in and out of focus of an image on a screen, resulting from buckling of the film either during exposure or high intensity projection.

BRIDGE

(1) Music—a transitional piece of music used to bridge a gap in a soundtrack or to connect one piece of music with another; bridge music (See also Sound Bridge); (2) action or dialogue created as a connecting link between one scene or action and another. Used mostly in comedy writing. In dramatic writing, "transition" is used more often (See also Transition); (3) a close-up or other angle inserted between two sections of the same scene and camera angle; (4) a narrow platform above a stage from which lights are mounted and upon which an operator may stand or sit to operate or adjust them. (See also Time Transition, Transition)

BRIDHAM'S NO. 1

See Frost.

BRIGHTNESS RANGE

The range of intensity of reflected light in the field of action, as measured by a light meter.

BRILLIANCE

The registered or perceived amount of lightness or darkness of a subject.

BROAD

A lighting instrument that casts a wide beam of soft light, used as a general fill light; single broads use one bulb, double broads use two.

BROAD RELEASE

A major release of a film in which it is distributed to hundreds of theaters across the country simultaneously.

BROADCASTING

Transmitting electromagnetic signals, usually beamed in all directions to a broad geographical area.

BRUTE

Trade name for a 225-amp arc lamp manufactured by Berkey-Colortron, Inc., used in cluster type quartz lights, referred to as "Mini Brute" or Maxi Brute."

BUCKLE TRIP
(Also Buckle Switch)
A safety device used on some cameras and projectors that stops the equipment in case of a jam.

BUCKLING
Bending or curling of the film edges resulting from shrinkage caused by dryness and/or tight winding.

BUDGET
The predetermined approved estimate of costs of a production, broken down into categories.

BUILDUP
The technique of using a variety of shots and creative editing to build suspense and action in a film.

BUILT-IN LIGHT METER
A light meter that has been manufactured into the camera and automatically selects the correct exposure. (See also Exposure Meter)

BULB
A lamp; the outer glass part of a light.

BULK ERASER
A degausser; an instrument that erases large rolls of magnetic tape or film by magnetically aligning all the iron oxide molecules. (See also Degausser)

BULLET HIT
A small explosive, hidden beneath a surface, used in special effects to simulate the impact made by a bullet.

BULLHORN
A hand-held electronic loudspeaker, frequently used by a director or one of the directing team, to call cues and instructions to extras and crew members on a location shoot.

BUMPER
(1) A period of time allowed between bookings of studio facilities to allow for equipment to be removed; (2) a program separator. (See also Program Separater)

BURNED OUT
(1) When an electrical component or lamp no longer functions because of excessive heat or normal life expiration; (2) a slang expression meaning that a performer, director, producer, or key industry executive has been either over-exposed or has lost the drive that once escalated him to the top.

BUSINESS
(Also Stage Business)
Action and bits of movement such as gestures and pantomime that emphasize the scene without dialogue, such as lighting a cigarette or pouring a drink.

BUSINESS AFFAIRS
The Business Affairs Department of a production company or network, which is responsible for negotiating contracts and for directing the financial activity and the day-to-day business activities generated by

a studio or major production company.

BUSY
Too much distracting, nonessential action, or too detailed or elaborate a setting.

BUTT SPLICE
A film splice in which two cut ends are joined by splicing tape without overlapping. (See also Butt-Weld Splice, Splicing, Straight Cut)

BUTT-WELD SPLICE
A film splice in which two cut ends are joined without overlapping by applying heat and pressure, as opposed to using splicing tape. (See also Butt Splice, Splicing)

BUTTERFLY
A large net, stretched on a frame, and rigged over an outdoor scene to diffuse and reduce light falling on the field of action.

BUZZ TRACK
A test film soundtrack used to adjust the lateral placement of an optical film in an optical sound reproduction system.

BUZZER
An alarm or signal indicating that shooting is about to commence, or that a shot has just ended.

B.W.
Abbreviation for black-and-white film.

C

CABLE
(1) Flexible insulated casing through which electrical or coaxial wire is able to transmit current safely or to broadcast a signal. (2) Abbreviated term for Cable TV.

CABLE MAN
(Also Sound Assistant)
The crew member who sets up the equipment for the boom operator, runs the cable from the set to the mixer, and wires the performers.

CABLE PULLER
In videotape production, the crew member who follows the camera operators to make sure that the cables are out of the way.

CABLECASTING
A system of transmitting programming produced exclusively for cable subscribers.

CABLE TELEVISION
(1) A television system that broadcasts signals by means of a coaxial cable. The signals can originate from one of the major U.S. networks or from original programming by an independent station or studio (such as MTV Music Television) or any of the pay television services offered to subscribers. Many cable systems such as Home Box Office and Showtime are available in major U.S. cities and are watched to supplement other channels. However, in many parts of the country, due to the terrain or the geographical distance from a major city, the cable is the only way that clear television signals from any source can be transmitted and therefore has become a necessity for television reception; (2) loosely refers to programming and services provided by a cable system: "I am watching the cable tonight"; "It will be on cable next month." (3) the coaxial cable itself. (See also Cable, Drop Cable, Grandfathering, Hub, Interconnection, ITFS, Penetration, Pole Rights, Public Access, SHO)

CABLE RELEASE
A device made of flexible wire used to activate a camera from a distance.

CABLELESS SYNC
See Cordless Sync.

CALL (Also Shooting Call)
The time, date, and location when each member of a production (cast and crew) is to appear for work.

CALL BACK
During a casting session, many performers are asked to read before the casting director, director, and producers. After each session, the

best of that group is asked to come back and read again. That performer is said to have a call back.

CALL SHEET

The announcement composed by the assistant director (AD) posted and updated as necessary. The call sheet is intended to be read by all involved in a production (cast, crew, staff) so that they will appear for work at the correct place and time.

CAM

Short for camera; especially used in hyphenated descriptive terms, such as mini-cams.

CAMEO (Also Cameo Performance)

A performance that is relatively short, but significant. Most cameo roles feature celebrities, which distinguishes a cameo appearance from a bit part. (See also Bit Part)

CAMEO LIGHTING

Lighting objects or performers in the foreground against a dark, solid tone background. (See also Cameo Staging)

CAMEO STAGING

Action filmed against a plain, often dark background; the key to this kind of shot is a nondistracting set or backdrop.

CAMERA

A light-tight box, capable of holding film and incorporating a lens, with a device for allowing light to enter the lens in a metered amount for exposing the film.

CAMERA ANGLE

The position of the camera in relation to what is to be filmed. (See also Eye Level Angle, High Angle Shot, Low Angle Shot, Normal Angle, Three-quarter Angle, Tilt Angle)

CAMERA AS NARRATOR

The technique of using the camera to convey information to the viewer that would otherwise have to be verbalized. Dialogue is replaced by visual representation, whether obvious (a ragged, skinny child gulping down a meal to tell of hunger), symbolic (a wilted rose seen by an old person who weeps, which perhaps tells a story of lost youth), or, literal (the camera becomes one's eyes, for example, if placed in a roller coaster, the viewer experiences the story of the ride without having to be told of the climbs and the falls).

CAMERA BODY

The central part of a camera that houses the fixed operative mechanisms, as opposed to detachable parts such as lenses, motors, or magazines.

CAMERA CAR

A vehicle (usually a car or truck) equipped to carry cameras and operators in order to film while the vehicle is in motion.

CAMERA DEPARTMENT

The personnel responsible for the maintenance and storage of all production cameras.

CAMERA IDENTIFICATION MARK

A brand name translated into a unique, usually geometric, trademark that is automatically exposed on the film, thus permanently recording on that film in which camera or model it was used.

CAMERA JACK

A connector designed to attach anything (such as power or sound) to a camera.

CAMERA LEFT/ CAMERA RIGHT

(1) Camera left literally refers to the left side of the camera as it looks at the subject. Camera right, is the right side of the camera. Basically created as a horizontal directional system to facilitate movement of the cameras, actors, or subjects to be photographed; (2) "Camera left" or "Camera right" represent verbal instructions to an actor or camera crew by the director, indicating which way they are to move during a shot.

CAMERA LENS TURRET

A television camera part to which several lenses can be attached. The turret revolves so that the lenses may be easily interchanged.

CAMERA LOG (Also Camera Report, Log, Shooting Log, Dope Sheet)

A printed form on which the camera crew records data pertinent to the production and the film shot on any particular reel. Blank spaces are provided for the following information: title, date, crew names, camera number, film emulsion number, shooting conditions, take numbers, filters. These logs are used by the editing department to help determine which shots will be incorporated into the workprint.

CAMERAMAN

See Cinematographer.

CAMERA MOUNT

A camera support system that allows easy camera manipulation for shots in which the camera moves (for example, panning or tilting). Mounts can be attached to tripods, booms, or dollies, or can be designed as shoulder braces.

CAMERA MOVEMENT

Movement of the camera while shooting, i.e., dollying, panning, craning, etc.

CAMERA NOISE

(1) The sound the camera makes while operating; (2) the unwanted sound of the camera heard in a sound recording.

CAMERA OPERATOR

The crew member who operates the camera under the supervision of the director of photography. It is the camera operator who sees the entire filmed sequence through the camera's viewfinder, while the director of photography generally stays in the background and supervises. On small shoots, however, it is not uncommon to find the director of photography acting as the camera operator.

CAMERA REPORT
See Camera Log.

CAMERA SPEED (Also Frame Rate, Projector Speed)
The speed at which film or tape is run through the camera, measured in frames per second or feet (or meters) per minute.

CAMERA TALK
The technique of photographing a subject as the subject looks directly toward the camera lens, typically used for interviews and news programs. (See also Talking Heads)

CAMERA TEST
(1) Routine check of the camera's operating functions, such as correct speed, steadiness, and focus; (2) the filmed audition of an actor for a role. (See also Audition, Sides, Test)

CAMERA TRACKS
See Dolly Tracks.

CAMERA USAGE
The use of a camera as predetermined by the director and cinematographer.

CAN
Any container that is used for safekeeping of film.

CAN, IN THE
(1) Footage that has been shot and is ready for processing; (2) film or tape that is ready for editing; (3) a completed production.

CANDELA (Also Candle, Standard Candle)
A term of measurement that refers to the amount of light reflected from any given surface.

CANNED LAUGHTER
See Laugh Track.

CANNED MUSIC
Recorded music, as opposed to music that is performed live.

CAPPING SHUTTER
A covering device used mostly on the lens of an animation camera to prevent the passage of unwanted light into the camera by closing automatically and independently of the set exposure shutter.

CAPSTAN
A spindle, usually with a roller pressing against it, that keeps magnetic tape flowing at a constant speed through a recorder.

CAPTION
Narrative or dialogue that is visually superimposed over a scene for descriptive or language translation purposes. Captions are often used to convey the time and location of the scene.

CARDIOID
(Also Shotgun Microphone)
A highly directional microphone that is more sensitive to sounds emanating from the area directly in front of it than from peripheral and rear areas. The long narrow pick-up pattern makes it particularly effective for outdoor shooting because it will block background noise. (See also Directional Microphone, Microphone)

CARS (Community Antenna Relay Service)

Also referred to as Cable Television Relay Service, the FCC-authorized microwave frequency band that allows relaying of air wave signals (broadcast signals) to cable television systems.

CARTOON

An animated film, usually of short duration, using drawings of the characters (cartoon characters) to act out a short plot or bits of connected action. (See also Animation)

CARTRIDGE (Also Cassette)

A completely packaged tape or film receptacle, fully self-contained, with two reels or spool devices feeding out and taking in the tape or film. A cartridge can be inserted into a recording or playback machine without manually threading the tape through the heads and then onto an independent take-up reel.

CARTRIDGE CAMERA

A camera designed to use a film cartridge (see also Cartridge), as opposed to utilizing independent take-up reels or spools.

CASSETTE

(1) In television, another word for cartridge (see also Videocassette); (2) in audio, loosely refers to a cartridge that is smaller than an 8-track cartridge.

CASSETTE RECORDER

An audio or visual recorder that uses only prepackaged cassettes or cartridges for recording and playback.

CAST

(1) Used as a noun, the performers selected to work in a production; (2) used as a verb, the process of selecting the actors for a production.

CASTING DIRECTOR

The individual whose responsibility it is to select various performers to fill acting roles in a production. Usually the lead roles are cast through a collaborative effort of the producer, director, and casting director. Often the stars have been decided upon before preproduction begins. Casting directors have varying degrees of authority, a factor decided by the producer and director. Most producers and directors want to be involved in the final approval of all casting except perhaps for nonspeaking roles, hence the casting director performs a suggesting and screening function offering suitable candidates for the director's and/or producer's final decision.

CATCHLIGHTS

The "sparkle" in the eyes of the subject being photographed. These tiny specks of light can be artificially created by placing small lights on or near the camera, which are also referred to as catchlights. (See also Eyelight)

CATTLE CALL

A little-used casting technique in which large numbers of actors and actresses are gathered for an audition. The term was coined because the process is reminiscent of a cattle drive.

CC FILTERS
A series of color-compensating filters (red, blue, green, yellow, magenta, and cyan) used to obtain precise color correction at the printing stage and sometimes while filming.

CEL
A transparent sheet, usually cellulose acetate, on which animation characters are drawn, with small changes on each sheet that will become movement when filmed in rapid sequence. Cels can also be used for opening and closing titles of a film by photographing through the transparent cel, when it is layered over artwork. (See also Animation)

CEL SANDWICH
Two or more cels layered so that a combination of artwork (background, characters, words, graphics) can be photographed and printed as one picture.

CELLULOID
The first widely used film for motion pictures. It was highly flammable (cellulose nitrate) and used mainly in the silent film era.

CELLULOSE ACETATE
(Also Cellulose Triacetate)
A clear, flexible film base used to make transparent sheets for animation and graphics drawings.

CEMENT SPLICER
A splicer that allows overlapping film pieces to be held securely while liquid cement is applied to seal the connection. (See also Splice)

CENTER OF ATTENTION
The subject on which the camera is focused.

CENTER OF PERSPECTIVE
The camera vantage point that coincides exactly with what the actor would see from that position. The camera becomes the actor's eyes, and the screen image projects what the actor would be seeing.

CENTER TRACK
A colloquial term that refers to the audio band manufactured on double perforation magnetic film; the track is a narrow strip positioned in the center of the film.

CENTURY STAND
See Gobo Stand.

CHALK OFF
See Blocking.

CHANGEOVER
The transfer of picture and sound from one projector to another. This occurs instantaneously with no loss of continuity.

CHANGEOVER CUE
A mark located in the upper right corner of a number of frames preceding the end of a reel, which alerts the projectionist that the changeover is due.

CHANGE PAGES
(Also Rewrites)
From the first time a cast and production staff meet to read the script, changes (rewrites) will be made. When they occur, script change pages must be issued. In the

early days of the feature film and television business, the industry-wide practice of color coding the change pages was initiated. A different color is assigned for each set of revisions. Although the order in which colors are used may change from one production company to another, the main colors used for the first five rewrites generally are yellow, pink, blue, green, and goldenrod. For example, a first rewrite would always be printed on blue paper, the second revision on pink, and so on. Whatever order of colors is decided upon is strictly followed, with copies of the change pages distributed to everyone involved with the production, including the director, cast, and crew.

CHANNEL

A narrow band of frequencies 6 MHz (megahertz) wide that carries a television signal.

CHARACTER

(1) A performer in a production; (2) a type of role in a film for which an unusual physical or personality type is wanted. (See also Stock Character)

CHARACTER ACTOR

An actor who specializes in playing especially distinctive roles, such as the villain or the tough guy, the little old man or woman, the snooty society type, or the young delinquent.

CHARACTER MAKEUP

Makeup that is used to change the appearance of a performer; that is, to make him or her look much older, younger, or to otherwise drastically alter his or her appearance. (See also Makeup)

CHARACTERIZATION

In performance, the distinctive traits of personality as interpreted by an actor from the work of the writer, with the guidance of the director.

CHASE FILM

A film whose plot is largely built around a chase or sequence of chases.

CHEAT

(1) The careful placement of people or props in front of the camera so that they appear casual and natural when photographed, but may actually be at unusual or awkward angles; (2) the photographing of shots, such as close-ups, against a background, other than that of the set, wild wall, or background against which the action is presumed to take place; (3) the removal of a fourth wall for the placement of cameras. The wall (see also Wild Wall) is then replaced when the cameras are moved to another angle that shows the wall in the background.

CHEAT THE LOOK

In order to show more facial reaction of a performer and yet maintain a profile shot, the director will instruct the performer to turn his/her face slightly more toward the camera. The actor would look awkward if viewed from a front or normal angle, but perfectly natural from the audience's point of view.

This enables the audience to see the performer's reaction while he still appears in profile through the lens.

CHECKERBOARD CUTTING
(Also Checkerboarding)

An editing technique in which A and B rolls are spliced to prevent the connecting point of the splice from appearing on duplicate prints by means of covering the overlapping film to the frame line with black leader. (See also A and B Rolls)

CHEMICAL FADE

A fade created chemically by immersing film in appropriate solutions (using either negatives or positives) so that the image is slowly darkened to blackout.

CHERRY PICKER

See Crane.

CHICKEN COOP

A boxlike light protected by a wire mesh screen to avoid anyone accidentally touching the bulb.

CHIEF RERECORDING MIXER

Directly responsible to the producer, his or her duties are to maintain maximum sound quality and perspective of the picture being rerecorded; supervise technicians involved in the mix; and make the combination of all tracks and/or units to the satisfaction of the producer.

CHILDREN'S FILM

A feature film, usually G-rated (see also MPAA Code) written and directed to entertain audiences under twelve years of age.

CHOREOGRAPHER

The individual who creates and stages dance sequences or a series of intricate movements in a production.

CHOREOGRAPHY

The design of dance, dancelike movements, or a prestaged complex series of physical movements (such as a fight sequence) for which exacting footwork, handwork, and body movement is required. For example, the fight scenes in *Rocky* required choreography as intricate in its design as the dance routines in *All That Jazz*.

CHRISTMAS TREE (Also Tree)

A metal structure on which several lights can be mounted.

CHROMA

The density of color.

CHROMATIC ABERRATION

A defect in a lens that causes the various colors in a beam of light to be focused at different points.

CHROMA KEY

A television matting technique in which two separately taped images appear on the same screen by superimposing one over the other. One subject is taped in front of a background made of special chroma key blue and is superimposed over action or with subjects or background that have been taped elsewhere. The blue background drops out resulting in the chroma-keyed subject and the other taped subject, or action, on one videotape; used often in taped special effects photography to place, for example, an actor wrestling with another actor (chroma key action) in a background of outer space with rockets and explosions (pretaped elsewhere).

CINCHING
Pulling the end of a reel of film in order to tighten it on the reel.

CINCH MARKS
Parallel scratches that can appear on film caused by trapped particles, or from improperly handled film that rubbed together during cinching.

CINEMATOGRAPHER
(Also Cameraman, Director of Photography)

Works closely with the director, supervises the camera crew, and is responsible for lighting and camera operation. In smaller productions, the cinematographer will operate the camera. The Academy of Motion Picture Arts and Sciences recognizes cinematography as a distinct creative function in the production of a film and accords an Oscar to the "Year's Best Cinematographer."
(See also Director of Photography)

CINE-
The prefix that denotes film or motion pictures.

CINE 8 CAMERA
(Also Regular 8 Film)

Older 8mm film with 8mm width and a perforation on one side at each frame line.

CINE CAMERA
A film camera for motion pictures.

CINEFLUOROGRAPHY
The process of filming X-rayed images on a fluoroscopic screen.

CINEMA
(1) Motion pictures or film, in general; "the movies." In Great Britain the term "cinema" is used instead of the "movies"; (2) a theater in which films or motion pictures are exhibited; (3) moving pictures with or without sound on flexible film.

CINEMA VERITE
A type of filming that originated during the 1950s in Europe. The filmmakers used portable equipment to conduct or recreate on-the-spot interviews while photographing the spontaneous action around the dialogue. This location filming resulted in a "film of truth" or reality in a way not possible in a controlled studio environment. (See also Direct Cinema, Documentary)

CINEMACROGRAPHY
The filming of tiny objects, such as insects and details of crystals. The most common means of such photography is through the use of plus-diopter lenses, lens extension tubes, or bellows extensions of lenses.

CINEMASCOPE
The process that uses anamorphic lenses to create a larger final screen image. With an anamorphic lens,

CINEMATOGRAPHY

the subject is vertically "squeezed" onto the film while the width of the frame of film remains normal. When the film is projected through equipment with a balancing lens, the vertical image is uncompressed, expands on the screen, and presents an undistorted picture that is larger than the image accomplished by filming through a regular lens. "Cinemascope" is the trade name for the process copyrighted by Twentieth Century-Fox; other companies followed Fox's lead and developed their own wide-image processes, such as Vista Vision, developed by Paramount.

CINEMATOGRAPHY

The photography of a motion picture as interpreted through the lenses of the cameras, camera angles, filters, and various photographic techniques used in filming.

CINEMICROGRAPHY
(Also Cinemicroscopy)

Photography through a microscope, thus enabling filming of extremely small objects, such as bacteria.

CINERAMA

A projection system using a large concave screen and 70mm anamorphic lens to create a three-dimensional quality.

CINEX PRINTER

A machine that prints the same shot with several different exposures on a "cinex sheet" in order to determine the best printer light for a shot.

CIRCLE OF BEST DEFINITION

The area within the circle of illumination that has the clearest image definition when photographed.

CIRCLE OF ILLUMINATION

The circular image created by a lens, a portion of which is called "circle of best definition," and is used for photography.

CIRCLED TAKES

The shots that are literally circled in pen or pencil on the camera log because they have been deemed acceptable (see also OK Takes) by the director and will be workprinted. (See also Workprint)

CIRCUIT DIAGRAM

A schematic with standard symbols to indicate the parts and component connections in electrical equipment.

CLAPBOARD (Also Clapper Board; Clapstick Board, Slate)

A small blackboard or slate with clapsticks at the top or bottom. On it is chalked all information pertinent to a shot; such as the film's name, date of the shooting, the director's name, camera operator, shot and take number, etc. The board is photographed at the beginning of each shot, which allows instant identification of the footage in editing. (See also Clapsticks, Electronic Clapper)

CLAPSTICKS

Sticks attached to a slate that, when clapped together, provide a visual and audio cue assisting the editor in synchronization of the film. (See also Slate, "Stick It," Upside-Down State)

CLARKE PROCESS

Action photographed through a diapositive plate, part of which has an image and part of which is left transparent for adding additional photography so that the image on the plate and added photography appear on the same piece of film.

CLASSIFICATION

The system of rating a film, a service currently provided by the Motion Picture Association of America, using a letter code to indicate a film's suitability for the viewing audience. Current rating codes are: G—general audiences; PG—parental guidance suggested; PG-13—Parents are strongly cautioned to give special guidance for attendance of children under 13. R—restricted, not open to persons under seventeen years of age unless accompanied by a parent; and X—no one under seventeen years admitted. (See also MPAA Code)

CLEAN ENTRANCE

The movement of an actor or prop from outside the action into the action being taped or filmed, as opposed to the camera panning toward the subject. Also, direction to the actor to enter the scene without creating sound or shadow before appearing in front of the camera.

CLEAN EXIT

The leaving of a scene by an actor or prop before the action ends and the camera stops shooting.

CLEANING

See Velveting Sound.

"CLEAR"

A word amplified or shouted across a set by the director's unit to indicate to all people not in the shot to leave or "clear the set," as the camera is ready for a take.

"CLEAR YOURSELF"

Verbal instructions from the director's unit to the performer to move so that the camera has a clear view of the actor; a view not blocked by a prop, a part of the set, or another actor.

CLEARANCE

The negotiated permission from the owner or his or her authorized agent of copyrighted material for the use of the copyrighted material in a production.

CLEARING BATH

A chemical bath that removes bleach or undeveloped silver from the film during processing.

CLICK STOPS

A system on the iris diaphragm ring of a lens assembly that allows the ring to "click" into place at each calibrated f-stop.

CLICK TRACK

Used to facilitate postrecording of a music soundtrack; the clicks are prerecorded in a tempo related to the film's action. The musicians (or conductor) listen to the clicks through headphones and match the rhythm as the soundtrack is recorded.

CLIENT
In the production of commercials or films produced other than for entertainment use (e.g., industrial or government films), the controlling contractor (individual or organization) who hires a film production company; the one for whom the film production company performs services.

CLIFFHANGER
An action film whose tension builds to a suspenseful climax. The term was introduced in silent film days when the last minute rescue of actors literally hanging from cliffs resolved the plot.

CLIP
(1) A brief, edited portion of a film, often used as a preview when an actor promotes the film on television (see also Preview, Trailer); (2) fasteners (plastic or metal) used to join rolls of films without splicing.

CLOSE SHOT
(Also Close-up Tight Shot)
A shot in which the head or the head and shoulders of the subject are all that is clearly visible in the camera's eye, or if not a person, a shot in which the camera is as close as possible to the subject. (See Extreme Close-up)

CLOSE-UP (CU)
Frame composition in which the camera's eye is on only a small part of the subject, that is, the head, or head and shoulders. (See also Close Shot, Shot, Skull)

CLOSE-UP LENS
(Also Diopter Lens, Plus Lens)
An additional lens that allows the regular lens to focus closer to the subject than usual; often used in cinemacrography.

CLOSING CREDITS
(Also End Titles)
Titles or credits that appear at the end of a film. (See also Credits, Titles)

COATED LENS
A lens coated with magnesium fluoride to reduce reflection and increase light transmission. (See also Lens)

COATING
Emulsions and oxides that are laid evenly on the film base. The process of applying the emulsions and oxides.

COAXIAL MAGAZINE
(Also Concentric Magazine)
A film (or tape) magazine whose supply and take-up reels lie opposite and parallel to each other.

COBWEB SPINNER
A device used to simulate cobwebs by blowing rubber cement toward the set. The rubber cement becomes long and stringlike, creating the effect of cobwebs.

CODE, THE
See Hays Office.

CODE OFFICE
See Production Code Office

CODED EDGE NUMBERS
An editing and general identification system whereby each roll of

film shot in a motion picture is marked on the edge with a series of numbers, which aid in identifying the rolls without viewing the footage and in editing synchronization. (See also Edge Numbers)

CODING MACHINE
(Also Edge Numbering Machine)
A device that prints numbers on processed film for ease in editing.

COLD
(1) Unrehearsed; (2) sound (music, voices, effects) heard alone, or "in clear." (See also In Clear)

COLD LIGHTS
Fluorescent lights.

COLOR
(1) Natural color photography, as compared to black and white; (2) anything added to a script/scene to contribute to or highlight authenticity or dramatic/comedic peak.

COLOR BALANCE
(1) The color distribution in the object to be photographed; (2) the response difference between outdoor and indoor color film. The difference is built into the film so as to correct for the difference in color intensities between daylight and indoor lighting.

COLOR BALANCING FILTER
A colored filter mounted in front of the light source that changes the color of the light, either lowering the temperature and creating a redder light, or raising the temperature, resulting in a bluer light. (See also Color Temperature)

COLOR-BLIND FILM
Film especially created to omit certain colors in printing. If the film is blind to yellow, yellow lights can be used when the film is processed without danger of coloring the film.

COLOR CAST
An undesirable tint, such as blue or pink, that appears on film by accident, however, it can be deliberately created for effect.

COLOR CHART
A test chart picturing colors of the spectrum. The Ilford Test Chart, for example, consists of two charts, one of which is composed of steps of deepening grays, with degrees of brightness parallel to the color half of the chart.

COLOR CORRECTION
Changing the color density of objects or images through the use of light filters, either with the camera or the printer.

COLOR DUPLICATE NEGATIVE
(Also Color Dupe Negative)
A color negative obtained in three ways: (1) The reversal process; (2) using black-and-white separation positives; (3) printing from a color master positive made from the color negative original. Optical effects can be added at this stage.

COLOR FILM
Film that contains one or more emulsions that when processed reproduce as different colors.

COLOR INTERNEGATIVE
A negative-image color print

made from a positive color original that is used for making release prints to protect the original. The color internegative is a final print complete with special or optical effects that were added during printing of the A and B rolls from the originals.

COLOR MASTER POSITIVE

A positive color print made from a negative color original used as an intermediate print to make 35mm color duplicate negatives and 16mm color and black and white duplicate negatives.

COLOR MATCH

Overall color consistency from shot to shot, reel to reel. Also a set of lenses used to create consistent color effect and reproduce balanced color images. (See also Color Balance)

COLOR NEGATIVE

Film from which a color master positive is made. Colors on the film are complementaries (see also Complementary Colors) of the subject shot; light areas appear dark in the negative and the dark areas light.

COLOR PRINT FILM

Special film used to print color positives from originals or color intermediates.

COLOR PROCESSING

The chemical treatment of film resulting in color images; the process is different from that used for black-and-white film.

COLOR REVERSAL FILM

This film has a positive color image when exposed in a camera and processed as opposed to a color negative.

COLOR REVERSAL INTERMEDIATE (CRI)

A film negative made directly from the original negative. Also called a one-Lite dupe, from which a balanced print can be made. Eastman Kodak developed the special stock needed to duplicate a film negative from the original film negative; it also allows all pertinent lighting and grading information to be transferred automatically onto the CRI by means of a computerized control process. (See also Grading)

COLOR SATURATION

The degree of color density present or absorbed through film.

COLOR SENSITIVITY

The degree to which photochemical emulsions respond to exposure of various wavelengths in the color spectrum.

COLOR SEPARATION

Three separate black-and-white negatives, one for each primary color (red, blue, green) prepared from an original subject or from a positive color film. Color prints can be made from these negatives by using cyan, magenta, or yellow filters in the printer. Color separation negatives are sometimes made of valuable color film for archival purposes, since these negatives have a long life and will not fade as do dyes used in color film. (See also Primary Colors, Three-color Process)

COLOR SEPARATION NEGATIVE

A black-and-white negative that

contains and has been processed for only one primary color.

COLOR TEMPERATURE
The measurement system created to catalogue and standardize the color of light sources by heating a temperature-controlled substance (can be a fragment of carbon), or black-body, until it begins to emit pure color. Temperature is measured in degrees Kelvin (same as degrees Centigrade except the Kelvin scale begins at minus 273 deg. Centigrade, and the Centigrade scale starts at 0 deg.). The lower the temperature, the redder the light, the higher, the bluer.

COLOR TEMPERATURE METER
A device designed to measure the the colors present in a film and indicates the additive printer light

COLOR-TRAN
A trade name for the autotransformer that increases voltage to standard lamps, in turn increasing their color temperature for effects in color photography.

COLOR VIDEO ANALYZER
A device that instantly assesses the colors present in a film and indicates the additive printer light intensities necessary to recreate a duplicate for normal results or for special effects.

COLORER
See Opaquer.

COLUMBIA BROADCASTING SYSTEM
CBS, one of the major U.S. Television networks. It has over two hundred affiliated stations.

COMBINED MOVE
The synchronized movement of a camera and actor during a shot.

"COME IN"
A phrase used by the director to have the camera move closer to the action.

COMEBACK
The successful return to work by a known actor, writer, producer, or director after a voluntary or involuntary (usually due to failure) absence.

COMEDY
Humorous action or dialogue that evokes mirth or laughter and/or creates a sense of levity. Also used to describe a film in which humor predominates. (See also Musical Comedy, Situation Comedy)

COMEDY OF MANNERS
Creating humor with dramatic characters by poking fun at their morals, etiquette, and social values.

COMETING
Small light spots that appear on a developed emulsion, caused by contamination in the processing bath.

COMMENTARY
(Also Narration)
Film narration created by an off-screen voice.

COMMERCIAL
(1) A film (usually between 10 and 57 seconds) created for the purposes of selling a product or creating an

attitude toward a subject; (2) highly saleable product; e.g., a film by George Lucas is more commercial than one by Ingmar Bergman.

COMMUNITY ANTENNA RELAY SERVICE
See Cars.

COMPILATION FILM
A film composed of a series of film clips, pieced together in editing and generally used for news purposes.

COMPLEMENTARY COLORS (Also Secondary Colors)
Colors produced by removal of the primary colors from the visible spectrum. Complementary colors are opposite one another on a color wheel. A colored filter will absorb its complementary colors: a red filter will absorb blue and green; a green filter will absorb red and blue; a yellow filter will absorb blue.

COMPLETION SERVICES
The additional services needed to complete a film after it is shot, such as processing, audio work, and special effects.

COMPOSITE MASTER
The completed original videotape from which copies can be made. Tape is always in a "positive" form, and unlike film, is not referred to as either positive or negative.

COMPOSITE MATTE SHOT
A double or multiple exposure created in the camera by using a different matte in each shot to create one picture when developed. (See also Matte Shot)

COMPOSITE PRINT (Also Married Print, Wedded Print)
(1) A film print in which sound and picture have been combined on the same strip of film; (2) a print made to create duplicate negatives that in turn will be used for making release prints.

COMPOSITION
The arrangement of all elements of a shot (people, props, point of view, light, color, etc.) to achieve the desired effect through the camera's eye.

COMPRESSION
A method of controlling the output of a signal, usually to prevent overmodulation in optical recording.

COMPUTER ANIMATION (Also Computer-Generated Animation, Computer Visuals)
Animation filmed or taped with the use of controlled images and patterns created by a special video computer.

CONCENTRIC MAGAZINE
See Coaxial Magazine.

CONDENSER, CAPACITOR
A wiring attachment capable of storing or partially blocking the flow of electrical current.

CONDENSING OPTICS
Lenses that narrow light beams from their sources.

COMPLEMENTARY MATTE
See Counter Matte.

CONE
A sound vibrator used in certain loudspeakers.

CONE LIGHT
A light that gives off a wide diffused beam.

CONFLICT
Dramatic action plot element in which two opposing forces struggle, such as characters vs. characters or characters vs. forces of nature.

CONFORMING
The editing process in which the original film is cut and matched to the edited workprint; also the separation of original film into A and B rolls. (See also A and B Rolls)

CONNECTOR
Two matching electrical parts that allow immediate joining or separating of electrical circuits.

CONSECUTIVE ACTION
Action as it is taped or filmed, not necessarily as it is written; many scenes are filmed or taped out of sequence. The action, as filmed or taped, is referred to as consecutive action.

CONSOLE
(Also Mixing Console)
A table, also called "the board," that controls the master mix of volume, filtration, equilization, and color intensity. Used for both audio and video tapes.

CONSTANT SCREEN DIRECTION
See Screen Direction.

CONTACT PRINTING
A technique in which an image from one piece of film can be transferred to another piece of film by sandwiching the two film strips together as they are exposed. Used especially in special effects films, such as the *Star Wars* series, in which the filmed spacecraft were later contact-printed onto the final film. (See also Step-Contact Printer)

CONTACT PRINTER
A film printer capable of keeping the film being printed and the stock on which it will be printed in simultaneous contact, with emulsions touching.

CONTAMINATION
The spoilage of photographic chemicals due to a foreign substance.

CONTINUITY
The binding of action and dialogue that ties a production together shot by shot, detail by detail, in a continuous, harmonious, and coherent flow.

CONTINUITY CUTTING
Standard editing technique in which film is pieced together in a smooth, connective story flow.

CONTINUITY LINK
A device, either tangible (props) or verbal, that helps to establish relationships and link together segments of a script.

CONTINUITY PERSON
See Script Person.

CONTINUITY SKETCHES
Drawings used during production by the director's unit as a guide to frame composition and shots.

CONTINUOUS ACTION
Uninterrupted capture of events, seen in a connective visual flow. Essentially this means that film time and real time will be the same during a sequence of continuous action.

CONTINUOUS CONTACT PRINTER
A printer that allows film movement in a steady, uninterrupted process.

CONTINUOUS EXPOSURE
Light that is not blocked or interrupted by a shutter in a camera.

CONTINUOUS OPTICAL PRINTER
An optical printer that moves the film through steadily and at an uninterrupted speed.

CONTINUOUS PRINTER
A unit (either optical or contact printer) designed to print film, taking it past the exposure aperture without interrupting the film speed.

CONTRACT
A written binding document specifying all terms of a deal in which services are to be rendered by the contractee to the production company, network, studio, or distributor, and agreed upon by all parties to the contract.

CONTRACT FILM
A film in which all negotiated and agreed upon expenses will be paid (backed) by a studio or sponsor, according to the agreement reached by the filmmakers and the backers.

CONTRACT NEGOTIATIONS
One of the first steps of production in which all business, financial, and shooting arrangements are reached with the production company and all who will be under contract, staff and crew as well as performers.

CONTRACT PLAYER
An actor who is paid a flat sum by agreement for services rendered for a specific part in one sequence of a production, as opposed to a performer on regular salary who appears throughout the entire production.

CONTRAPUNTAL SOUND
Sound created to play against or contrast with the action, usually music, as opposed to sound that adds plausible, predictable, or smooth background.

CONTRAST
The range between the lightest and darkest tones in a screen image. (See also Gamma)

CONTRAST FILTER
A colored filter, most commonly yellow, amber, or red, that increases

the brightness of objects in black-and-white film. Objects the same color as the filter become lighter gray; objects in tones complementary to the filter color become deeper gray. (See also Filter)

CONTRAST GLASS
See Viewing Filter.

CONTRAST RANGE
The brightness range between the lightest and darkest areas of a set, and the ability of film to record both ends of the spectrum of light and dark in the scene.

CONTRASTING SCREEN DIRECTION
See Screen Direction.

CONTRASTY
In black-and-white filming, a dominance of extreme light and dark tones with little balance in gray tones; in color, extreme brightness and intensity of colors.

CONTROL BOOTH (Also Booth)
Found in television production, a room set apart from the stage in which the technical director sits in front of his control board, allowing him to switch from one camera to the other at the director's instructions. A bank of monitors is mounted on the wall facing the director, assistant director, P.A., and technical director enabling them to see exactly what each individual camera sees. Oftentimes, a partitioned-off area is included in the booth, from which special visitors to the set can watch the show being taped without interfering with the production process. In episodic television, usually the producer, associate producer, and writers, while in the booth, will follow along in their scripts, making notations of where changes should be made.

CONTROL PANEL
A box or table with easy access, displaying electrical controls, for use in presetting lights or sound (or sound equipment).

CONTROL SIGNAL (Also Control Track, Crystal Sync)
An electronic beat recorded on the magnetic soundtrack that keeps the sound in synchronization with the picture. The beat is produced by a crystal in the audio recorder or a generator in the camera.

CONVERSION FILTER
A colored filter mounted in front of the lens that allows daylight-balanced color film to be used with incandescent light or indoor film in daylight.

COOKIE (Also Cuke, Cukaloris, Kukaloris)
An opaque attachment for light sources in which patterns, such as snowflakes, have been cut. As the light beams through the cookie, the pattern is reproduced as a shadow on the background or field of action.

COOL IMAGE
The use of blue tones in lighting or printing to tone down or "cool" the red tones. (See also Warm Image)

COPTER MOUNT
Camera mount used in aerial or helicopter photography that attaches the camera to the helicopter,

as opposed to the camera being hand-held by the operator. The mount assures accurate focus and eliminates vibration.

COPYRIGHT
The legal ownership of a created work to insure payment when the work is duplicated, played, or performed. To be legally copyrighted a work must be registered with the U.S. Register of Copyrights. A small circled letter c (©) appearing near the title indicates the work is copyrighted.

CORDLESS SYNC
(Also Cableless Sync)
Any of various systems available to insure synchronization between camera and sound recorder, without the use of a sync-pulse cable.

CORE
The plastic, wood, or metal centerpiece around which film is wrapped.

CORE-TO-CORE
The process of rewinding film from one core to another.

COSMETIC MAKEUP
Natural looking makeup that is used to enhance an actor's appearance. (See also Character Makeup, Makeup)

COSTUME DESIGNER
The person who designs and creates all costumes in a production.

COSTUMER
The individual who secures, handles, and fits clothes worn during a film.

COUNTER
(1) A dial with visible numbers that keeps track of film length in feet, half-feet, or frames; (2) a slight movement by an actor to create space, mood, or better reveal business or reaction.

COUNTER MATTE
(Also Complementary Matte)
Matte used during the second exposure of an in-camera matte shot to prevent double exposure of previously exposed portions. (See also Matte)

COVER
(1) To film a scene from all possible angles to insure that the action will be shown in a satisfactory and revealing manner when the shots are edited together; (2) to film an event, as for news purposes (see also Coverage); (3) to set up a camera position that will include all desired action.

COVER SHOT (Also Establishing Shot, Master Shot)
A long shot that covers all the action in one scene for the purposes of establishing or reestablishing action, location, and mood. (See also Reestablishing Shot)

COVERAGE
(1) Filming of an event, usually for television news; (2) the air time received from such filming; (3) the number of close-ups or other angles shot by the director in addition to his master scene. (See also Break It Up)

COVERING
A staging disaster whereby one actor blocks another from camera view, or stands in his key light. Usually prevented by the director.

COVERING POWER
The capacity of a lens to pick up and produce both vertically and horizontally a clear image over a given film frame size.

CRABBING
Irregular dolly movement accomplished only when all of the crab dolly wheels can be steered simultaneously.

CRAB DOLLY
Camera-mounting support with wheels that can be steered in any direction, usually with an adjustable-height camera seat that raises to approximately five feet for higher shots.

CRANE (Also Cherry Picker)
A moveable vehicle with a long, rotating, high-rising arm on which a camera can be mounted.

CRANE CREW
The people responsible for the operation of a camera crane.

CRANE OPERATOR
The person who drives the vehicle or participates in the movement of the camera.

CRANE SHOT
A shot with the kind of movement available only with a crane. This would be an extremely high-low shot (see also Craning). Also, a shot made with the camera mounted on a crane. (See also Boom Shot, Shot)

CRANING
Raising or lowering a camera that is mounted on a crane.

CRAWL (Also Crawling Title(s), Roll-up Titles, Creeper Titles)
(1) Titles and credits that appear on the screen moving upward from the bottom to the top; (2) the device that produces titles.

CREDITS
A list that appears at the beginning and often also, in a somewhat different form, at the end of a film with the names of all involved in the production. Usually the more important credits, such as the producer, director, and stars are listed in the opening credits. (See also Billing, Opening Credits, Closing Credits, Starring, Titles)

CREEPER TITLES, MAIN TITLE
See Crawl.

CREW CALL
A posted notice to crew members stating time and place of shooting; a rough schedule is usually available at the beginning of production, however, updated sheets are posted and must be checked daily.

CREW MEMBERS
See Grips.

CRI
Abbreviation for Color Reversal Intermediate.

CRITICAL FOCUS
Perfect, exact focus on the subject of choice, objects in front of or behind that point are usually blurred to varying degrees, depending on the depth of field and the lens used. (See also Focus, Follow Focus, Reflex Focusing)

CRITICS
Those employed by the television, radio, and print news media for the purpose of evaluating productions for the audience. Things such as content, cast, direction, technical values, writing, and overall appeal are reviewed. (See also Review, Success d'Estime)

CROP
To cut out some of the area seen in a shot either with the camera before the shot is taken, or by blocking the undesirable area of the negative so it will not appear in the finished print. (See also Television Cutoff) Cutoff)

CROSS-CUTTING
(Also Intercutting)
The rapid cutting back and forth between two (or more) different scenes so that it appears that the action in all scenes involved is taking place simultaneously.

CROSS BACKLIGHT
See Kicker Light.

CROSS DISSOLVE
See Cross Fade.

CROSS FADE (Also Cross Dissolve, Segue)
To fade out one sound or picture gradually while simultaneously fading in another. While the intensity of one decreases, the intensity of the other increases. (See also Sound Dissolve)

CROSS LIGHT
(Also Cross Lighting)
(1) Light that falls on a subject at angles to the lens axis; (2) the intentional use of such lighting.

CROSS MODULATION
Audio distortion that arises from the optical recording of sound by the variable-area method. If uncorrected, the result is an unwanted overemphasis of sibilant high frequency tones.

CROSS MODULATION TESTS
Tests used to avoid cross modulation by correcting, in advance, negative-positive densities for a variable-area optical soundtrack.

CROSS-OWNERSHIP
The ownership of two or more types of communication media by the same business or individual. Television stations and telephone companies are prohibited by the FCC from owning cable systems in their service area. Television networks are prohibited from owning cable systems anywhere in the United States.

CROSSING THE LINE
A director's mistake that results in the actors appearing from the audience's point of view to look or move in a direction other than that established in the preceding shot of

the same scene. (See also Axis, Imaginary Line)

CRYSTAL CHECKER
A tool to determine if the crystal control on a camera or recorder is operating properly.

CRYSTAL MICROPHONE (Also Piezoelectric Microphone)
A cableless microphone in which the sound vibrates from a crystalline substance that in turn generates a small amount of voltage that can be used as an audio signal. (See also Microphone)

CRYSTAL MOTOR
A motor in a camera whose speed is governed by the vibration of a crystal.

CRYSTAL SYNC (Also Control Signal, Crystal Control, Crystal Sound, Crystal-Controlled Motor, Crystal-Controlled Sync, Crystal Cordless Sync)
The synchronization of camera picture and sound by means of an accurate internal crystal-controlled timing device. This highly precise system does not require connecting cables between the camera and the sound recorder.

CU
Abbreviation for Close-Up.

CUE
(1) The action, dialogue, gesture, or off-stage indication for the actor to start his/her action immediately (see also Voice Cue); (2) any mark, symbol, system, or mechanism that indicates the beginning or ending of action, speech, music, effects, or edit point.

CUEPRINT
A positive projection print containing cues for postrecording.

CUE CARDS (Also Flip Cards, Idiot Cards, Idiot Sheets)
Large cards upon which dialogue or narration is written in easily read hand-written letters. The cards are held out of camera range to be read by the performer. Even though they are sometimes called "idiot cards" or "idiot sheets," the cards are used frequently by top professionals. In addition, refers to the various mechanical teleprompting devices in use, especially on television news shows. (See also Teleprompter)

CUE IN
To begin action, narration, sound effect, special effect, or music. The term is also used by audio engineers when a photograph record is set, or cued, to begin at a predetermined point.

CUE LIGHT
A small light switched on and off by a technician to indicate when a narrator should begin to speak.

CUE MARK
A mark or hole to indicate the starting place of the film, magnetic video, or audio tape for editing, interlock projection, or recording as a system of starting in sync.

CUE PATCH
A special self-adhesive magnetic

or metallic substance that when placed on the edge of film, initiates a printer light change or an automatic projector stop.

CUE SHEET (Also Dubbing Cue Sheet, Mixing Sheet)

Instructions prepared by one production technician for another. The sheets detail what has been done to the tape or film and make it apparent what is to be done or added, including proper levels, settings, and any special data; especially used by the sound mixer as prepared by the editor. (See also Music Cue Sheet)

CUKE, CUKALORIS
See Cookie.

CULT FILM

A film that appeals to only a certain portion of the general public but which is regarded by that portion with such great and long-lasting enthusiasm that the film achieves fame amongst its followers. Some popular examples include: *The Rocky Horror Picture Show* and *Night of the Living Dead.*

CURL

The bending of a film's edges caused by changes in humidity.

CURVATURE OF FIELD

Focusing the lens on an image in which the focal points are on a curved rather than flat plane.

CUT

(1) "Cut"—the director's command to stop the camera and/or action; (2) the instantaneous switch from one shot to another (See also Invisible Cut); (3) the point of joining two pieces of cut film; (4) to eliminate a shot, scene, sound, or effect from a film; (5) to shorten a scene through editing. (See also Action Cutting)

CUTAWAY (Also Insert Shot, Reaction Shot)

(1) A break from the main action to another shot; (2) cut to a close-up whether within the main action or not, but definitely to something not directly involved with the featured action; (3) an insert shot to show related details or reactions that comment on the principal action; (4) an editing technique to maintain continuity and/or establish editing pace or creativity. (See also Shot, Switchback)

CUT BACK

The switch back and forth from two parallel action sequences; a return to action that took place before a cutaway.

CUT TO

To change instantaneously from one camera or one shot to another.

CUTOUT ANIMATION

Figures and forms cut from flat materials that are jointed for movement during filming. (See also Animation)

CUTTER

See Editor, Film Editor.

CYAN

A blue-green color.

CYC STRIP

A lighting device comprised of several lamps in a line that spread an even light over a cyclorama.

CYCLE

A period in which many feature films of a similar type or theme are released by several major studios and/or independents, such as space films capitalizing on *Star Wars* popularity, westerns, or gangster films.

CYCLE ANIMATION

A series of several cels drawn to show repetitive motion (such as running or water flowing) that are filmed over and over in series sequence to show continuous action. (See also Animation)

CYCLORAMA

A large curved backdrop at the back of a set or stage, usually plain, so that colored lights or props can be used without changing backdrops.

D

DB
(1) Abbreviation for Delayed Broadcast; (2) abbreviation for Decibel.

DGA
Abbreviation for Director's Guild of America

D-MAX
The density of the totally unexposed portions of emulsion in a reversal film after developing and fixation.

D-MIN
The density of the totally exposed portions of emulsion after developing and fixation.

DAILIES (Also Rushes)
The first print, made immediately from the day's original footage, which is viewed on a daily basis by the director and involved production staff. (See also Syncing Dailies)

DAILY PRODUCTION REPORT
See Production Report.

DARK END
The area surrounding the film processor where film is fed into it without light being able to strike the film.

DATA RINGS
The guide rings on lenses that indicate point of focus, f-stop, and depth of field.

DAY
In a script or shooting schedule, the indication that filming is to be done in daylight, or in light created artificially to simulate daylight.

DAY EXTERIORS
Filming outdoors on location in daylight or light created to simulate outside daylight.

DAY FOR NIGHT
Shooting a scene during the day and lighting it so that it looks as if it were shot during the night. (See also Night for Day, Night for Night)

DAY OUT OF DAYS
A breakdown of all the actors and the days they work. Drawn as a chart, this form will indicate which actors will work on which day.

DAY PLAYER
A performer who is hired on a daily basis.

DAYLIGHT
(1) Natural outdoor sunlight or indoor light during the day (See also Sunlight); (2) artificial light created to simulate daylight; (3) a color film

designation indicating film balanced by the manufacturer for daylight filming.

DAYLIGHT COLOR FILM
Color film balanced by the manufacturer to give good color rendition when shot in daylight. (See also Daylight)

DAYLIGHT CONVERSION FILTER (Also Daylight Filter)
A camera lens filter that, in effect, makes color film balanced for artificial light respond as daylight film. (See also Daylight, Daylight Color Film, Filter)

DAYLIGHT LOADING
Cameras and related equipment designed to permit the use of daylight-loading spools.

DAYLIGHT LOADING SPOOL
A reel fitted with light-tight flanges situated so the film is completely protected when loading or unloading in moderate light.

DAYTIME
The television programming hours from 10:00 A.M. to 4:30 P.M. Eastern, Mountain, and Pacific Time Zones, and from 9:00 A.M. to 3:30 P.M. in the Central Time Zone. (See also Early Fringe Time, Family Hour, Fringe Time, Late Fringe Time, Prime Access, Prime Time)

DAYTIME PROGRAMMING
Television shows created or designated for broadcast during daytime, especially soap operas, game shows, talk shows, and syndicated reruns of prime time network shows.

DEAD
An acoustical term referring to an enclosed space in which noise or reverberation is reduced to a very low level; an enclosure in which there is a large amount of sound absorption.

DEAD SPOT
(1) A point in production where the action slows or dulls so that a rewrite or other production technique might be called upon to enliven it; (2) a position or place on a set or stage where there will be no action during a shot; (3) a part of a set not seen by the cameras.

DEAD SYNC
(Also Editorial Sync)
A term used in editing to indicate exact sound and picture synchronization.

DEAL LETTER
(Also Deal Memo)
A written declaration of a tentative agreement between individuals or companies, outlining the terms of a deal to be included in a forthcoming contract. Deal memos (or letters) are commonly used in many phases of the industry, for example, between film or television production companies and talent, including performers, writers, and directors. A deal memo is an immediate way to express a genuine intent to sign a person or project, and is a low-cost, somewhat binding preliminary to the expense and time-

consuming effort of preparing contracts.

DEBLOOPING
See Bloop.

DECIBEL (Also DB)
(1) A unit of measure of sound. The higher the decibel range, the louder the sound; (2) the unit of power-level difference, a measure of the response in electrical communication.

DECOR
The furnishings used to decorate and dress the set of a production.

DECORATIVE PROPERTIES
Props that create the atmosphere specified by the director. These props are mainly for decoration and are rarely handled by the performers. (See also Action Props, Prop, Practical)

DEEP FOCUS (Also Deep-Field Cinematography, Deep-Focus Cinematography)
A film style in which everything in the frame, near and far, is kept in sharp focus by using small f-stops and/or short focal length lenses. (See also Focus)

DEFINITION
(1) The sharpness or clarity with which objects or individuals are photographed; (2) the fidelity and clarity of sound reproduction; (3) the capability of a film emulsion to separate fine detail.

DEFOCUS
To allow the action intentionally to go out of focus during a shot by focusing on a point closer to the camera, thereby reducing the depth of field. Usually a lens with longer-than-normal focal length is required. (See also Focus)

DEGAUSSER
An instrument used to erase magnetic tapes and film, or to demagnetize recording heads. (See also Bulk Eraser, Demagnetize)

DEGRADATION
The loss of any sharpness or clarity due to duplication.

DEGREES KELVIN
A unit of heat measurement using the Kelvin scale, commonly used to measure lamp and color temperatures.

DELAYED BROADCAST (DB)
The local market broadcast of a previously aired network program that has not yet aired in that market, but was recorded and kept for future broadcast by the local station.

DEMAGNETIZE
Loosely, to erase or degauss a magnetic recording tape or the heads on a recorder. (See also Degausser)

DEMO
A demonstration tape, record, or script used for an audition or to display one's talents.

DEMOGRAPHICS
The statistical composition of an audience, including age, sex, and viewing habits.

DENOUEMENT
The unraveling or resolution of a story, after the climax, which ties loose ends and explains any outstanding points in the plot.

DENSITOMETER
An instrument used to measure photometric density.

DENSITY
(1) The most applicable term to describe the light-stopping characteristic of film, which is the light-stopping power of opaque silver deposits in the processed photographic emulsion; (2) the degree of darkness present in a negative determined by the amount of opaque silver deposits on the film; (3) the brightness of light in a scene. (See also Opacity)

DEPTH OF FIELD
The point at which objects are in sharp focus, which is determined by the f-stop used and the focal length of the lens.

DEPTH-OF-FIELD SCALE
A device that measures the depth of field for a specific lens at a specific distance and aperture setting.

DEPTH-OF-FIELD TABLE
A table that lists the depth of field for a specific lens at various points of focus and aperture.

DEPTH OF FOCUS
A term often mistakenly used for depth of field, depth of focus has little application in cinematography and is a photographic term indicating how far a film can be moved behind a lens. Motion picture film is held in a fixed position from the lens by the gate. (See also Gate)

DERIVED SOUND
The combined sound of two stereo tracks, replayed on a third, middle, loudspeaker.

DESATURATION
(1) The elimination of color from film, resulting in a monochrome effect; (2) a more common use of the term, the effect created by such color removal.

DETAIL SHOT
A close shot of a small object, part of an object, or a very close shot of a person, perhaps showing just the eyes.

DEUCE
A colloquial term for a 2,000-watt lamp.

DEUS EX MACHINA
(God from the Machine)
A plot device or gimmick that provides an artificial solution to a problem. The term is derived from Greek and Roman drama in which a god would appear at the play's climax to help the characters.

DEVELOPER
Any chemical that reacts with exposed silver or light sensitive material in film so as to bring out a photographed image in film.

DEVELOPING
The chemical process whereby photographed images appear on film. The first stages of film processing, as opposed to the end stages of fixation and drying.

DEVELOPMENT

(1) A creative development department, existing in most major studios and production companies, in which fragments of ideas for future projects or story lines are discussed, fleshed out, and selected for script assignment and possible production; (2) an important change in a character or plot direction.

DGA

Abbreviation for Directors Guild of America.

DIAGONAL

(1) The diagonal distance across the image area of film in a camera; (2) the distance in current use to measure the area on a television screen.

DIAGONAL CUT
(Also Diagonal Splice)

The most common method of cutting a splice. On magnetic tape and film this minimizes splice noise.

DIAGONAL WIPE

See Wipe.

DIALOGUE

The words in a script, as spoken by the actors in a production; speech as opposed to music or sound effects. (See also Empty Dialogue, Wild Lines)

DIALOGUE COACH

The individual who assists the performer in learning proper delivery of his or her lines as the character in a production, especially when an accent is being used.

DIALOGUE REPLACEMENT

The technique of looping or dubbing new dialogue to replace unsatisfactory dialogue. This is accomplished in a dubbing studio, with the actors timing their delivery against the scenes already shot.

DIALOGUE TRACK

A soundtrack of lip-synched dialogue, as opposed to music or sound-effect tracks.

DIALOGUE TRIANGLE

A writing structure in which the audience is the third person, as listeners in a dialogue between two people whose lines have been written to convey specific information to the viewer.

DIAPHRAGM

An iris in a lens or spotlight. This iris controls the amount of light that enters through the lens. (See also Iris)

DIAPHRAGM PRESETTING

Selecting an f-stop that will give the desired depth of field, and then lighting the scene of action to the level that will provide the desired exposure.

DIACHROIC FILTER

(1) A filter used on lamps to create the effect of daylight by reflecting excessive red and transmitting colored light; (3) a filter used in color printers to absorb certain colors while allowing others to be printed on the film. (See also Filter)
color printers to absorb certain colors while allowing others to be printed on the film. (See also Filter)

DIEGESIS

The flow of events in a story narrative in the film's space and time, as opposed to the time it would take for them to occur in reality. For example, a complete boxing match as was shown in the *Rocky* trilogy may take approximately fifteen minutes on the screen but significantly longer in reality. (See also Dramatic Time, Expansion of Time, Film Structure, Filmic Time and Space)

DIFFERENTIAL FOCUS
(Also Split Focus)

Focusing at a compromising point between two subjects in order for both of them to appear equally sharp in the depth of field range.

DIFFERENTIAL REWIND

An instrument that allows simultaneous winding of film on multiple reels.

DIFFUSED LIGHT

Light emanating from a large source and scattered so that it casts either no shadow or a soft shadow.

DIFFUSERS

Devices, such as muslin, fine nets, fiberglass, gauze, or grooved glass, used in front of a lens or lamps in such a way as to soften the image to be photographed by reducing the harshness of the lighting. Diffusers can make skin appear younger. (See also Umbrella)

DIFFUSION LENS

An auxiliary lens used as a diffuser when mounted in front of the lens being used, to soften the image to be photographed. (See also Diffuser, Lens)

DIFFUSION SCREEN

A translucent screen, spun glass, or similar material used to reduce harshness in lighting when placed in front of the light source. (See also Diffuser)

DIMMER

A commonly used term for rheostat. An electric device that reduces the amount of light emanating from one or more lights.

DIMMER BANK

A group of rheostats, or dimmers, that allows for the dimming or brightening of a group of lights. These units can be permanent or portable.

DINKY INKY (Also Inky Dink)

A small, low wattage spotlight, "Dinky" for small, "Inky" for incandescent.

DIOPTER LENS

An auxiliary attachment to a regular lens that allows close-up filming of the subject. (See also Close-Up Lens, Lens)

DIRECT CINEMA

A term derived by Albert Maysles to distinguish this type of "close observation" filmmaking from cinema verité. Direct cinema is on-location, nonfiction filming with portable cameras and sound recorders, used to record sound and action as it actually happens as opposed to a recreation of events. (See also Cinema Verité, Documentary)

DIRECT COLOR PRINT
A color print made in one step directly from original color film.

DIRECT COSTS
Actual costs directly related to a specific production as opposed to indirect overhead costs. (See also Production Overhead)

DIRECT CUT
The abrupt change from one camera shot to another camera shot.

DIRECTION OF LOOK
The performer's line of sight as a shot ends; the direction in which a performer is looking.

DIRECTIONAL MICROPHONE
A mike that has a greater sensitivity to sound from one direction than from others. It is used when trying to avoid picking up sound that is not part of the main action. (See also Cardioid Microphone, Microphone)

DIRECTIONAL PATTERN
The directional areas of sensitivity as pick-up patterns of a microphone. Basic pick-up patterns include nondirectional, which picks up sounds from all directions; bidirectional, which has a two-area pick-up pattern or stereo pick-up; and unidirectional, which picks up sounds coming basically from one direction.

DIRECTOR
The individual who is responsible for the total look of the film, combining and supervising all the creative elements, including performances, camerawork, and editing. The director interprets the script and is in charge of getting it on the screen in accordance with either his or the producer's creative vision and intent.

DIRECTOR OF PHOTOGRAPHY
See Cinematographer.

The individual responsible for lighting and camera operation on a production. The director of photography works closely with the director and supervises the camera crew. When filming a commercial, the director of photography will often operate the camera.

DIRECTOR'S CUT
The director's edit of the film; this is the first cut following the editor's rough assembly of the film in sequence, which is also done to the director's specifications. All DGA represented directors are entitled to a director's cut under the terms of the Guild's basic agreement.

DIRECTOR'S FINDER
A small viewfinder, usually worn by a director around the neck, that is calibrated to show lens focal length, used to view the field of action in order to select the proper lens focal length to be used on the shot. (See also Viewfinder)

DIRECTORS GUILD OF AMERICA (DGA)
A guild representing the collective bargaining and professional

interests and welfare of directors, assistant directors, and stage managers, with offices located in Los Angeles and New York City. (See also Guilds)

DIRECTOR'S NOTES
Comments on and criticisms of the show (the performances, script, camera work, etc.) given by the director after a reading, rehearsal, or taping.

DIRECTORY, THE
See Academy Players Directory.

DIS
Abbreviation for Dissolve.

DISCONTINUITY
A visual clash or mismatch of action, props, set furnishings, lighting, or wardrobe as one shot ends and the next shot begins; a jump cut.

DISCOVERED
(1) A word used to express that an object or actor is on the scene when the shot begins; (2) a commonly used term when someone is found by a talent scout, agent, or producer who intends to give that person a chance at stardom.

DISCOVERY PAN (Also Discovery Dolly Shot, Discovery Zoom)
A shot (pan, dolly, or zoom) that introduces something to the audience that was not revealed to them when the shot began.

DISJOINTED NARRATIVE
A film in which the story line is a series of nonsequential events.

DISSOLVE (Also Lap Dissolve)
One of the most common optical effects used to connect two scenes, accomplished by overlapping a fade out with a fade in, often with a momentary superimposition of the two shots on the screen, appearing to melt into one another, until the first shot fades completely out and the second shot has faded in. (See also Cross Fade, Fade, Soft Cut, Sound Dissolve)

DISSOLVE ANIMATION
A technique that allows gradual changes in artwork or subjects by using closely spaced dissolves. (See also Animation)

DISSOLVE IN
To fade in with a dissolve to a new picture. (See also Dissolve, Fade)

DISSOLVE LAPSE
Short shots connected by dissolves that record processes similar to time lapse photography.

DISSOLVE OUT
To fade out with a dissolve. (See also Dissolve, Fade)

DISTANT SIGNALS
Television signals originating from a point too far away to be received by ordinary home equipment, that cable systems offer to subscribers on a limited basis, as controlled and specified by the FCC.

DISTORTION
(1) Any optical distortion or malformation of the image on a tape or film (positive or negative) whether caused by equipment, technical error, power supply, or faulty film or tape stock; (2) any unwanted change or distortion in the sound of an audio track.

DISTORTION OPTICS
Specific optical attachments that create special image effects when used on a camera or optical printer.

DISTRIBUTION
The rental, sale, or lease of feature films for exhibition. The distribution of television programs is referred to as syndication.

DISTRIBUTOR
The company or organization that distributes feature films by rental, sales, or lease to commercial outlets. Major distributors sometimes are involved in the production of the films they market.

DITTY BAG
A small catch all sack, usually made of canvas and hung open below a tripod, containing scissors, tape, and other small items needed by a camera crew.

DMA (DESIGNATED MARKET AREA)
A Neilsen classification assigned to an area in which the local television stations are watched more frequently than stations located in other areas, but whose signals reach the local market. (See also Neilsen, Arbitron)

DOCUMENTARY
A nonfiction story or series of events usually made in actual locations with nonperformers; a record of events as they happened or as they affected the lives of those who experienced them. A factual account of a subject. (See also Direct Cinema)

DOLBY
Trade name for a noise-reduction system that can be included in the original film and, with the aid of special equipment, can be reproduced in a theater. Dolby is designed to enhance the fidelity of the sound and is capable of reproducing stereo sound.

DOLLY
(1) To move the camera on a dolly while filming; whether toward or away from the subject, or following behind or preceding a moving subject; (2) a wheeled vehicle with a platform that carries the camera and camera operator during shooting, often with a boom mike attached and room for the boom operator.

DOLLY IN
Camera movement toward the subject.

DOLLY OUT
Camera movement away from the subject.

DOLLY PUSHER
A camera crew member who physically moves the dolly during set up or shooting.

DOLLY SHOT
Any shot that is accomplished by moving the camera around the set on a dolly. (See also Shot)

DOLLY TRACKS
(Also Camera Tracks)
Level channel rails upon which a dolly is mounted in order to control and smooth movement during a shot.

DOPE SHEET
A list of the shots on a roll of film and their descriptions, especially used for shots taken for a documentary or when filming unscripted shots. (See also Camera Log)

DOT
A small, round device used for forming shadows. (See also Finger, Flag, Gobo, Scrim, Target)

DOUBLE
A performer who takes the place of another performer, usually a principal cast member, during dangerous stunt shots and location shots in which the performer is not seen in close-up. In addition, doubles usually sit in for a star during lengthy setup shots and light checks, which are not done with cameras and sound rolling. (See also Stand-in)

DOUBLE CARDIOID
See Figure Eight.

DOUBLE CHAIN
The simultaneous running of A and B rolls on two television projectors in order to use cutaway shots during a television interview.

DOUBLE 8 MILLIMETER FILM
Sixteen-millimeter film perforated like the older 8mm format; half of the emulsion area is exposed, then flipped over in the camera to expose the other half. After processing, the film is slit and spliced together for a continuous projection.

DOUBLE EXPOSURE
(Also Superimpose)
A picture made by exposing the same piece of film more than once, either in the same camera or in an optical printer. A double exposure is film exposed twice to two separate scenes or images; however, reexposure can occur as many times as desired. The superimposed images may be placed in any visual relationship such as side by side, or individually vignetted.

DOUBLE FEATURE
Two feature films presented in a theater for the price of one.

DOUBLE-HEADED PROJECTION
A separate picture track and soundtrack projected synchronously in interlock on a double-system projector.

DOUBLE KEY
See X Lighting.

DOUBLE PERFORATION STOCK (Also Double Perforated Stock)
Film manufactured with perforations on both edges.

DOUBLE PRINT TITLES
The superimposition of titles over

DOUBLE SCRIM

other images through A/B roll printing.

DOUBLE SCRIM
Two layers of scrim, used for extra darkening.

DOUBLE SYSTEM
A camera and sound system in which the sound is recorded on a tape deck separate from the camera, as opposed to a single system. (See also Interlock System, Single System)

DOUBLE-SYSTEM PRINT
A workprint in which the available soundtrack is separated from the film and run on a single strand magnetic track, which through the use of a special interlocking projection system synchronizes sound and picture; used in the early stages of postproduction to preview a film.

DOUBLING
A situation in which one actor plays more than one part in the same production.

DOWN
(1) A directive or script indication to diminish the volume of a musical or sound effects track; (2) playing down a role or portion of a role, as opposed to overacting; (3) a tonal quality assumed by an actor to differentiate between dialogue and narration, when both are required in the same speech.

DOWN SHOT
A high-angle shot taken with the camera looking down on the subject.
(See also High Angle Shot, Shot)

DOWN TIME
The time during which equipment on a set is not in use due to repair or servicing of equipment.

DOWSER
An instrument that cuts off the beam of light from a projector when changing reels from one projector to another.

DRAFT
Various attempts at writing a script from the first, or rough, draft to the end resulting script, or final draft.

DRAMATIC IRONY
Often used to create suspense by giving certain knowledge of relationships or events to the audience, but not to the characters; a technique especially used in soap operas.

DRAMATIC TIME
(Also Script Time)
The time period during which the action takes place, which could be days, months, or years, as distinguished from the time on the screen, or production time, which may be fifty minutes to two and a half hours in which to depict the script's time period. (See also Diegesis, Expansion of Time, Film Structure, Filmic Time and Space)

DREAM BALLOON
An animated bubble or cloudlike image, sometimes appearing near the performer's head, used for visualization of a dream or thought sequence.

DREAM MODE
Shots that show that the action is

the imagination, daydream, or dream of the character.

DRESS
To prepare the set for the day's shooting; to get ready for a performance.

DRESS REHEARSAL
A type of rehearsal in which the cast is in costume, the set is dressed, and a formal rehearsal takes place. The cameras and sound recorders are usually off.

DRESS SHOW
In television, the first of two tapings of a show that are taped in a single day. The first show is considered a type of dress rehearsal, although it is taped before a live audience. The second show is called the air show. The best scenes of the two shows will be combined in editing to form the final broadcast show.

DRIVE MOTOR
(1) The motor that powers the camera and sound recorder in sync; (2) any motor that powers the main mechanism of a camera, projector, or recorder as opposed to a nondrive motor such as a fan motor or crystal motor.

DRIVE-IN THEATER
An outdoor area for film exhibition in which the audience pays, drives in, and watches from their cars. Each parking place is equipped with a small loudspeaker that fits in the car window.

DROP
See Backdrop.

DROP CABLE
The coaxial cable used to connect the cable network to an individual home or building. (See also Cable)

DROP SHADOWS
Shadows that appear to the side and below title letters.

DRUM TITLER
An instrument that makes motion picture titles from printed copy, by placing the vertical strip of copy around the drum, and photographing the copy while the drum rotates. The end result is screen titles that move upward on the screen.

DRY BLOCK
A stage rehearsal without cameras.

DRY BOX
The last unit in a film processing machine in which dry warm air circulates around developed film.

DRY EDIT (Also Paper Edit)
An editing technique during which the film or tape is viewed by the director, editor, and producer who decide at that point what edits are to be made. They are not, however, actually accomplished during that editing session.

DRY END
That end of the film processor at which the dry box is located.

DRY RUN
A stage of rehearsal of the actors and/or camera movement without actually filming or taping.

DRYING

The final step in film processing in which dry warm air is circulated around developed film in a dry box. (See also Dry Box)

DUB

(1) A copy of a magnetic video or especially audio tape; similar to a dupe of a film print; (2) the process of duplicating a video or audio tape (see also Dubbing); (3) an actor lip-synching or dubbing in lines, whether repeating lines already spoken (see also Looping) or dubbing an entire film by replacing one language with another, as done with foreign films.

DUBBER

An audio playback machine for use with magnetic or optical recordings. (See also Mixer)

DUBBING

(1) Another term for recording. The process of making a duplicate of a video, but especially audio, tape; (2) the process of replacing original voices on a soundtrack as done when a foreign language picture is dubbed into English; (3) also, loosely, to mix soundtracks, or rerecord several soundtracks onto one track, or postsynchronizing voices to match action. (See also Dub, Looping, Mixing, Postsynchronization)

DUBBING CUE SHEET

See Cue Sheet.

DUBBING SESSION

Studio time scheduled for looping or dubbing in which actors record lines of dialogue as they watch a workprint of the action. The dubbing or looping may be in a different language from that originally filmed with the action. (See also Dub, Dubbing, Looping)

DUBBING STAGE
(Also Looping Stage)

A small audio studio specifically set up for looping and dubbing.

DULLING SPRAY

A spray-on, nonhardening liquid, usually an aerosol, sprayed on bright or shiny surfaces to reduce their highlights or reflection.

DUNNING-POMEROY SELF-MATTING PROCESS

A technique in which an actor moves in front of a blue background; the action is shot with a bipack camera containing a bleached and dyed print of the desired background action and panchromatic negative film. The actor becomes a matte against the blue light and the appropriate background from the bipack print is recorded onto the negative.

DUPE

(1) A duplicate print or copy of a film, videotape, or sound recording (see also Dub); (2) specifically, to make a duplicate negative from a positive (see also Picture Duplicate Negative); (3) the command to make a duplicate or dub.

DUPLICATE NEGATIVE

A negative made from a master positive or positive original, that is used for making release prints. (See also Picture Duplicate Negative)

DUTCH TILT ANGLE
(Also Dutch Angle)
A camera angle accomplished by tilting the camera from its normal vertical and horizontal axes.

DUTCHMAN
A strip of fabric that is glued over the joining edges of flats and painted or decorated to match, thus hiding the joining crack.

DYNALENS
Trade name for a commonly used gyro-controlled optical system that reduces erratic camera movement and vibration in shots made from moving vehicles or with telephoto lenses.

DYNAMIC RANGE
The difference in volume between the loudest and the softest sounds in a recording or live audio pick-up.

EARLY FRINGE TIME

One of the designated television time periods for the purpose of selling advertising and gathering ratings data. For Eastern, Mountain, and Pacific Zones, the hours are from 4:30 P.M. until 7:30 P.M. Central Time Zone's early fringe time is from 3:30 until 6:30 P.M. (See also Daytime, Family Hour, Fringe Time, Late Fringe Time, Prime Access, Prime Time)

EAST

In animation direction, the right side of the cel, field chart, or animation table. (See also Animation)

ECHO

A reverberation of sound that can occur intentionally or by accident due to technical problems in recording dialogue, music, or sound effects.

ECHO CHAMBER

A room or device that adds reverberation to sound.

ECU (Also XCU)

Abbreviation for the term extreme close-up.

EDGE FOGGING

Accidental exposure of the edges of film. Caused by misuse of film cans, light leaks in the film magazines, or in the camera itself.

EDGE NUMBERING MACHINE

(Also Coding Machine)

A machine that prints edge numbers on processed film and magnetic sound tracks. These are printed as opposed to latent-image, edge numbers. (See also Edge Numbers)

EDGE NUMBERS

The standard editing system that allows for easy synchronization of sound and picture, and for synchronization of the workprint with the original footage during the conforming stage. Four- to six-digit numbers appear at six- to twelve-inch intervals along the edge of the film (original, workprint, and lip synch sound film) by one of two processes: latent-image or printed. Latent-image edge numbers are exposed by the film's manufacturer and appear on each roll in development; these numbers are transferred to subsequent rolls by a printer. Printed edge numbers are assigned during production and are printed by an edge-numbering machine so that all picture and sound rolls have the same sequence.

EDGE STRIPE

The magnetic sound track strip that appears on the edge of film so manufactured. (See also Edge Track)

EDGE TRACK
The magnetic film soundtrack manufactured as a narrow band along the edge of single perforation magnetic film, opposite the perforations.

EDIT
To cut and assemble creatively all filmed, taped, and produced elements of a production for presentation to an audience, including adding and omitting footage (see also Edit Out) and synchronizing the various sound tracks with the picture.

EDIT OUT
To remove portions of a film, videotape, or sound track during the editing phase of postproduction.

EDITED MUSIC TRACK
Music or sound effects that are precisely edited to accompany specific action on the screen, as opposed to laid-in track.

EDITING
The process of cutting, splicing, and assembling the film and soundtrack (or videotape) for the final viewing print. (See also Edit)

EDITING MACHINE
A device consisting of a vertical or horizontal viewing apparatus capable of running films and sound tracks separately or in sync for the purpose of editing.

EDITING TABLE
A flat table upon which are mounted some of the pieces of equipment needed to edit motion picture film and soundtracks, such as rewinders, splicers, and sound readers.

EDITOR
(Also Film Editor, Cutter)
The individual who edits film, videotape, or soundtracks.

EDITORIAL PROCESS
(Also Editing Process)
All tasks related to the assemblage of the final footage for release, including cutting, post production, duplication, and synchronization of picture, sound tracks, and special effects.

EDITORIAL SYNC
(Also Edit Sync; Dead Sync)
A system of arranging the picture and sound rolls in such a way that they pass through the equipment with the soundtrack directly opposite its corresponding frame of picture.

EDUCATIONAL FILM
(1) A film written and produced especially to teach a skill or impart facts or concepts. Many such films are shown exclusively in classrooms (See also Instructional Film, Training Film; (2) loosely, any film that results in public awareness of facts, values, or issues.

EE
Abbreviation for Electronic Editing.

EECO NUMBER
See SMPTE.

EFFECT FILTER
A camera lens filter that produces

a special visual effect, such as fog or color change. (See also Filter)

EFFECT MACHINE
A projection device for magnifying and projecting moving background effects.

EFFECTS (FX)
A term encompassing various kinds of effects, such as sound effects, optical effects, special effects, or all of the preceding.

EFFECTS PROJECTOR
(Also Effects Machine)
A slide or film projector that projects special effects images onto a backdrop or translucent screen.

EFFECTS TRACK
Refers to magnetic audio tape; a separate track, or loosely, a separate tape, which contains sound effects only, as opposed to dialogue or music.

EI
Abbreviation for Exposure Index.

8MM FILM
Motion picture film that is eight millimeters wide. Regular 8 or cine 8, an older type of film, has the same size perforations as sixteen millimeter, however, twice as many (80 perforations) per foot. These perforations are found at the frame line. Super 8, the newer type and more frequently used today, has 72 perforations per foot, located opposite the middle of the frame. Eight-millimeter film is the most commonly used film in home movie cameras.

ELECTRIC EYE
See Photoelectric Cell.

ELECTRIC DRIVE MOTOR
Any synchronous or variable-speed electric motor that drives a camera, recorder, or projector.

ELECTRICIAN
In the industry sense, a person who works with the functions of set and stage lighting as opposed to a licensed installation electrician or contractor. (See also Gaffer)

ELECTRICIAN'S KNOT
A simple knot in which stage cables are tied at the connector joint so that they will not become separated by accidental strain or pull.

ELECTRO-PRINTING
A process whereby the master magnetic soundtrack is transferred directly to the release print without the use of an optical soundtrack printing master.

ELECTRONIC CLAPPER
(Also Electronic Slate, Automatic Slate)
An electronic method of synchronizing picture and sound for workprints by which a device in the camera sets a tone corresponding with the picture. (See also Clapboard)

ELECTRONIC COMPOSITE
(Also Electroprint)
A release print with composite sound transferred from the magnetic soundtrack directly onto the print instead of from an intermediate optical soundtrack printing master.

ELECTRONIC EDITING (EE)
A method of editing videotape in which the unwanted part of the tape is cut out electronically with the use of a production console, as opposed to physically slicing the tape.

ELEVATION DRAWING
A drawing to scale of a set's vertical components depicting all shapes and dimensions of the set.

ELEVATOR SHOT
A shot in which the camera is moved vertically or straight up and down, as opposed to moving the camera in an arc as with a crane shot. (See also Shot)

ELLIPSOIDAL SPOTLIGHT
A light, used on sets, that casts an elliptical beam.

ELLIPTICAL EDITING
(Also Elliptical Cutting)
A film editing technique that eliminates much action. Such a technique often uses jump cuts.

ELS
Abbreviation for Extreme Long Shot.

EMERGENCY LIGHTS
Stage, set, or house lights controlled from the front of the house for use in the event of a power loss backstage; loosely referred to as panic lights.

EMMY
The award given out annually by the Academy of Television Arts and Sciences. Starting in 1949, the gold-tone statue of a woman with wings holding a globe, which represents the Muse of Art holding an atom of Science, has been given for outstanding achievement in many categories, such as performances and technical achievements. The name Emmy comes from the words "image orthicon tube." The award was originally named "Immy," but a typographical error resulted in the name "Emmy."

EMOTE
An acting directive in which emotion is expressed instantly for the camera.

EMPTY DIALOGUE
Speech in a script that is not essential to plot or character revelation. Empty dialogue is the first to be cleaned out of a script if time becomes a consideration.

EMULSION
(Also Film Emulsion)
A light-sensitive coating that is spread over or bonded to a film base. The three main film emulsions are: (1) for unprocessed film, gelatin and silver salts; (2) for processed film, gelatin and metallic silver; and (3) for magnetic sound, iron oxide. (See also Base)

EMULSION BATCH
A specific emulsion mix coded by number on each roll of film and its box by the manufacturer. The number allows for purchase of many rolls of film from the same emulsion batch as well as being a reference point for defective film.

EMULSION NUMBER
The manufacturer's number used

EMULSION PILEUP

to identify a film's emulsion batch. (See also Batch Number, Emulsion Batch)

EMULSION PILEUP

A distortion caused by a lump of emulsion in the gate of a camera, projector, or viewer.

EMULSION POSITION
(Also Emulsion Side)

The side of the film on which the emulsion must be for right-reading or correct visual image. If the emulsion is toward the screen, the film has a B-wind projection position; if toward the projection lamp, the film has an A-wind projection position. If the emulsion position is reversed, images will also appear wrong-reading or reversed. (See also A-Wind; B-Wind, Right Reading; Wrong Reading)

END TEST

A test of exposed footage at the end of a roll of film to evaluate the film before the remainder of the roll is processed.

END TITLES

See Closing Credits.

ENLARGEMENT PRINTING

An optical printing technique that results in a larger image produced from a smaller frame, such as 16mm to 35mm.

ENTERTAINMENT FILM

Commercially produced and distributed film properties for mass markets in theaters or on television. (See also Educational Film)

ENTRANCE

(1) The arrival to the scene of action of a performer, animal, robot, or moveable prop; (2) the doorway, opening, or area through which the performer or subject enters the scene.

ENVIRONMENTAL SOUND

Spontaneous sound (usually low-level) from the field of action that is picked up by the mikes.

EPISODE

One show or installment of a serialized television program; one complete television show. A season is usually comprised of twenty-four new episodes, which are sometimes rerun between seasons or in syndication.

EPISODIC

(1) A description of a type of television programming in which the same characters appear in each show and whose plot(s) are based on their exploits. "All in the Family" is an example of episodic television, as opposed to television news shows or movies of the week; (2) a type of film story in which the main action is in many locations as opposed to a single location; (3) a series of productions written around a person or persons' adventures or experiences. In that sense the *Star Wars* trilogy is episodic.

EQ

The abbreviation for "equalize" or "equalization." Sound technicians are often heard to say a recording has to be EQ'd for proper sound balance.

EQUALIZE
(Also Equalization)

The electronic changing or altering of audio frequencies to improve the quality of a recording. Equalization is often used to remove extremely high or low frequencies and can improve the quality of recorded speech and the balance between vocals and music. (See also Balance)

EQUALIZER

The equipment used to equalize sound recording. Equalizers are widely used in sound mixing and are usually built into modern production facilities.

EQUITY

See Actor's Equity Association.

EQUITY THEATER

A theater for stage productions with ninety-nine seats or less.

EQUITY WAIVER

A stage production in which certain Equity union restrictions have been waived so that production companies with very small budgets will be able to stage a show.

ERASING

Removal of sound, picture, or both from a magnetic track by passing the tape or film through a high frequency magnetic field. The magnetic track can then be reused. (See also Degausser)

ERECT IMAGE

A visual image that appears in correct vertical position as the eye sees it, with the top side up.

ERECTING SYSTEM

The system of lenses or prisms that allows an erect image to be seen through the camera's viewfinder. In old cameras without the erecting system, the image appears upside-down. (See also Erect Image)

ESTABLISHING SHOT
(Also Cover Shot)

A shot that, at a glance, depicts key elements, spatial relationships, and details of time, place, and characters in a scene. Usually a long shot at the beginning of a scene, however, medium shots and even close-ups that contain identifying elements are also used. Sometimes called a cover shot, since this shot covers or establishes important elements of the action or scene for the viewer. (See also Reestablishing Shot, Shot)

ESTIMATES

The approximate calculations of production costs based on the requirements of a proposed script. Budgets are put together based on the estimates.

ETHNIC FILM

A film genre in which the plot, characters, casting, and setting depicts, or has special interest to, a particular race, religion, or cultural group.

EXCITER LAMP

A key element in optical sound reproduction, it excites and activates the current modulated by an optical soundtrack as the track moves between the lamp and the phototube resulting in the sound on the film track being amplified through loudspeakers and heard in synchronization with the action.

EXCLUSIVITY
(1) The contracted right for the exclusive services, usually by a studio or production company, of the contractee, such as a performer or writer; (2) an often-used clause in a television station's syndication contract granting exclusive presentation rights of a program or film in a specific broadcast area. The FCC prevents cable stations from broadcasting distant signals of shows that would violate a local station's exclusivity contract.

EXECUTIVE PRODUCER
The studio or production company executive who supervises the over-all production of a film or television show, including the work of the producers and director, but one who usually does not handle the day-to-day duties of producing the film. In television production, however, the executive producer often becomes more involved on a daily basis.

EXHIBITION
The showing of films in commercial theater outlets.

EXHIBITOR
A theater owner, manager, or sponsor of films for exhibition.

EXIT
(1) The leaving from the scene of action of a performer, animal, robot, or moveable prop; (2) the doorway, opening, or area through which the performer or subject leaves the scene.

EXIT LINE
The last line spoken as, or just before, a performer exits. Sometimes called a tag line.

EXPANDED CINEMA
Films that are produced using the latest technical equipment, such as computer visuals or holograms.

EXPANSION OF TIME
Expanding the actual time it would take to accomplish an action into "screen time," used to create suspense, for example. This is accomplished by filming or taping the same action with multiple cameras or by editing cut-aways to repeat the action from a different or more detailed angle. If a character reaches for a gun, the actual time to grab it may be three seconds; this can be expanded by the camera first showing the gun, then the face of the actor, then the gun again, then the hand reaching for it, then the fingers clasping it, all of this takes significantly longer on the screen than in reality. (See also Diegesis, Dramatic Time, Film Structure, Filmic Time and Space)

EXPERIMENTAL FILM
An independently produced film, usually noncommercial, that reflects the filmmaker's personal vision in technique or story line; usually expressing a unique and sometimes bizarre artistic viewpoint. (See also Underground Film)

EXPLOITATION FILM
A film that emphasizes, and takes

full advantage of, a subject such as sex, violence, or horror that will attract a specific audience; often a pejorative term. (See also Blacksploitation Film, Sexploitation Film)

EXPOSITION
The story technique that tells the viewer at the beginning of the story who the characters are, their part in the story, and where the story takes place.

EXPOSURE
(1) The process of allowing light to touch film (through a camera) for a controlled length of time, which results in an image appearing on the film after the film is processed; (2) the length of time the film is exposed; also (3) the amount of light used in the exposure; (4) the amount of publicity released regarding a specific performer, motion picture, television show, or property; (5) the audience recognition factor itself; for example, Archie Bunker has received an enormous amount of exposure in the American television market; (6) a general feeling created by publicity, such as bad exposure or good exposure. It is good exposure for a film or TV show if the star discusses it on a nationally televised broadcast.

EXPOSURE CALCULATOR
A guide with a moveable disk that when positioned for certain light conditions indicates probable best exposure.

EXPOSURE GUIDE
A printed guide indicating probable best exposures for a particular film under various lighting conditions.

EXPOSURE INDEX (EI)
The technical term for film speed; the manufacturer's numbers, such as ASA ratings, that correspond to a particular film's sensitivity to light, or emulsion speed, when exposed and processed within specific conditions. The numbers enable correct determination of exposure setting. (See also ASA Speed, Exposure)

EXPOSURE LATITUDE
The degree of under- or overexposure to which a film can be subjected and still produce acceptable results.

EXPOSURE METER
(Also Light Meter)
A unit manufactured to measure the amount of light upon or reflected from the subject in order to set the camera for correct film exposure. There are many such meters, from simple hand-held units to more complex photoelectric and automatic devices that attach to the camera. Measuring the intensity of light is critical to assure that the film will not be under- or overexposed. (See also Built-in Light Meter, Incident Light Meter, Meter Reading, Spot Brightness Meter)

EXT
Abbreviation for "exterior."

EXTENDED CONTROL
A switchboard designed so that one or more circuits can be switched on or off from a remote position.

EXTENSION TUBE
A tube that allows extension of the lens focal length thereby increasing its magnification; the tube adapts between the camera and the lens.

EXTERIOR
(1) Any shot filmed outside the stage; (2) an indication in a script that a shot is to be made outdoors; (3) a shot made outdoors or made indoors to look as if it were shot outdoors. (See also Studio Exteriors)

EXTERIOR LIGHTING
Lighting for outdoor shots, either natural light, artificial light, or a combination.

EXTERIOR SOUNDS
(1) The sounds of trains, planes, and traffic as they are heard outdoors; (2) sounds that occur in outdoor locations, whether created naturally on location or artificially in the studio; (3) spontaneous sounds as they occur on outdoor locations.

EXTERNAL VIEWFINDER
See Viewfinder

EXTRAS
Actors hired for background or to create reality or atmosphere on a set, such as customers in a restaurant. Usually extras have nonspeaking roles and receive no screen credit; however, if extras are playing the part of an excited or angry crowd, they may be directed to shout, cheer, or chant. Most extras in major productions perform through SEG (Screen Extras Guild).

EXTRAPOLATION OF THEME
A production in which the central theme is significantly changed, deepened, or transformed by other related themes.

EXTREME CLOSE-UP
(Also ECU, XCU)
A detailed close shot of an actor or subject, such as eyes, a dewdrop on a flower, a hand, or a candle flame. (See also Close Shot, Shot)

EXTREME HIGH ANGLE SHOT
A shot made from a high angle by placing the camera well above the subject and shooting down on it. This can be accomplished with the camera attached to or atop such diverse objects as a crane, a building, a mountain, or an aircraft. (See also Shot)

EXTREME LONG SHOT
(Also ELS, XLS)
Used mostly on exterior shots, a view taken a great distance from the subject. If there are actors in the shot they would normally appear very small in comparison to the surroundings. (See also Shot)

EXTREME LOW-ANGLE SHOT
A shot made from a low angle, by placing the camera far below the subject and shooting up toward it. (See also Low Angle Shot, Shot)

EYE CONTACT
A shot that gives the impression that the subject is looking at the

audience, because the subject is looking directly into the camera.

EYE LEVEL ANGLE
A shot taken with the camera lens placed at the performer's eye level. (See also Camera Angle, Shot)

EYELIGHT
A tiny light placed near the camera that does not register exposure value, but is used to put highlights, or bring out the sparkle, in a performer's eyes. (See also Catchlights)

EYEPIECE LENS
The lens at the end of the camera's viewfinder through which the camera operator looks. (See also Lens)

F-STOP
The numbers on a camera lens that represent the speed of the lens at any given diaphragm setting and indicate the amount of light that can pass through the lens. The number is the result of dividing the focal length of a lens by the diameter of the lens opening. The smaller the f-number, the larger the opening and the dimmer the light needed to shoot. The larger the f-number, the smaller the opening and the brighter the light needed to shoot. (See also Stop, T-Stop)

F-STOP BAND
(Also F-Stop Ring)
A part of the lens barrel that, when turned, sets the iris at any of the various f-numbers marked on the ring.

F-SYSTEM
A system that calibrates lens diaphragms in terms of exposure.

FADE (FI, FO)
(1) An optical effect in which the picture gradually disappears into blackness (fade-out); or the gradual appearance of the picture from a black screen (fade-in); (2) the gradual loss of sound (fade-out); the gradual emergence of sound (fade-in). (See also Printer Faders, Variable Opening Shutter)

FADE SCALE
(1) A device with a series of marks and a pointer attached to the fader mechanism, used to insure smoothness in the increase or reduction of exposure (See also Fade); (2) any device on a camera or printer that creates fades; (3) in audio use, a device that permits gradual attenuation of sound.

FADER
(1) A device on a printer and camera that makes fades by slowly increasing or decreasing the amount of light available to the film; (2) also, the control used to adjust the level of sound to produce fades.

FALL-OFF
(1) The gradual reduction of light falling on a set by the use of flags, barndoors, or other such devices; (2) the weakening of light as its source is moved to a greater distance.

FALSE MOVE
An incorrect movement or action by the performer.

FALSE REVERSE
A reverse angle shot that, when the edited shots are projected, results in the subject or performer moving or looking in the wrong direction due to too great a change

in camera position during the shot. (See also Shot)

FAMILY HOUR
A casual term for the early evening television viewing hour from 7:00 to 8:00 P.M., including Fringe Time and part of Prime Time. (See also Daytime, Early Fringe Time, Fringe Time, Prime Access, Prime Time)

FAN
(1) An enthusiastic admirer or supporter of a performer; (2) a unit to create high wind. (See also Wind Machine)

FANTASY
A film that bears little relation to reality as we see it in the world around us and that, in its imaginative divergence from reality, often has much in common with fairy tales and science fiction stories.

FAR SHOT
A long shot.

FARCE
Broad comedy that borders on the ridiculous: comedy with greatly exaggerated characters and actions to create laughter.

FAST
(1) Films with higher than average sensitivity to light; films that do not need as much light in which to film a normal exposure; (2) lenses with their maximum f-stop numbers near one.

FAST CUTTING
(Also Flash Cutting)
Editing that uses many short shots.

FAST FILM
Highly light-sensitive film with an exposure index of at least one hundred.

FAST LENS
A lens with a larger maximum aperture that allows more light to pass through to the film. (See also Lens)

FAST MOTION
Footage that has been shot at a rate slower than normal but projected at normal speed, which results in the action appearing to move comically fast. (See also Single-Frame Shooting, Stop Motion, Time Lapse Photography, Undercrank)

FAVOR
To choose camera angles and/or lighting that spotlights and flatters a performer.

FAY LIGHTS
A type of 650-watt bulb used mainly with Molefay lights for even illumination, especially in outdoor filming.

FCC
Abbreviation for Federal Communications Commission. Established as part of the Federal Communications Act of 1934, the FCC is a federal agency that regulates all television and radio broadcasting in the United States and territorial waters. The FCC issues rules, and regulates practices in such areas as station ownership, call letters, and I.D. regulations, amount of commercials, and the information to be kept in a daily log by each station.

FEATURE
(1) To introduce or present a relatively new performer in a leading role; (2) a screen credit for a leading role that is billed under the starring credits; (3) a feature film.

FEATURE FILM
A film, either fictional or based on fact, over an hour in length and made to be released and screened in commercial movie theaters.

FEDERAL COMMUNICATIONS COMMISSION
See FCC.

FEED
(1) The mechanism on a camera or printer that guides the film into the machine; (2) the transmission of a broadcast signal, for example, a network transmission to an affiliated station for airing. (See also Affiliate)

FEED LINES
To read lines from off stage in order to assist the actor in front of the camera. This technique is used if lines are forgotten by an actor; but more often is used as a way of assisting a solo actor in delivering his or her lines more believably by giving him or her someone to respond to.

FEED REEL
The reel by which film or tape is pulled through equipment.

FEEDBACK
(1) A loud, electronic screech caused by two live pick-up units getting too close to one another; (2) an electronic circuit arranged so that part of the output is returned to the circuit early to improve performance or stability of that circuit.

FEEDER CABLES
(Also Feeder Lines)
Coaxial cables connecting the main trunk line that carries the broadcast signal from the cable company to the individual cables that are hooked up to subscribers' homes.

FEET PER MINUTE
A term relating to standard film speed. The speed of silent film is 60 feet per minute, sound film speed is 90 feet per minute, and slow motion photography is 180 feet or more per minute.

FG
Abbreviation for Foreground.

FI
Abbreviation for Fade-In.

FIDELITY
Accuracy of reproduction of sound and color.

FIELD CAMERA
A portable, lightweight camera used in location filming.

FIELD GUIDE
A chart of the areas covered by various ranges of camera positions.

FIELD OF ACTION
The view in front of the camera within the limits of the focal length and the distance to the subject. (See also Action Field)

FIGURE EIGHT
(Also Double Cardioid)

A mike that is more sensitive to sounds either directly in front or behind it, than to those coming from the sides.

FILL LEADER

Leader, usually yellow or white, used to fill in picture blanks in a workprint or to fill parts of a sound roll between sound sections, especially when separate rolls are prepared for different sounds. (See also Leader)

FILL LIGHT (Also Fill, Filler, Filler Light, Fill-In Light)

(1) A light usually placed near the camera to fill in shadows caused by the brighter key light; (2) any light, whether from a luminaire or a reflector that comes from the direction of the camera and is pointed toward the action to balance or remove shadows; (3) a weaker light used to balance the shadow side of a subject. (See also Contrast Range, Light, Lighting Ratio)

FILL PROGRAMMING
(Fill Program)

A program of brief duration used between the end of one broadcast and the beginning of another; used especially at the end of a movie or sports presentation on television in which the end does not occur on an exact half hour and something must be added to "fill" the time until the next scheduled program.

FILM

(1) The art of motion pictures in general; a feature film; (2) to photograph a motion picture; (3) a strip of transparent, flexible material, usually cellulose acetate or triacetate, that is perforated and coated with a light-sensitive emulsion, and iron oxide for film with a magnetic sound track. Standard motion picture film sizes are 8mm, super 8mm, 16mm, 35mm, and 70mm.

FILM ARCHIVE

A film library and repository for printed materials connected with the films.

FILM ART

Using cinematography to express creatively human experience.

FILM BASE

The usually transparent, flexible support surface on which photographic emulsions and magnetic coatings are placed.

FILM CAPACITY

The amount of film that can be used by a film camera, a projector, or related equipment.

FILM CEMENT

A term used commonly, but erroneously, to refer to the welding solvent used in film splicing.

FILM CHAIN (Also Telecine)

The mechanism by which film is projected into a television system.

FILM CHAMBER

A light-tight container (camera film box or film magazine) that houses both unexposed and exposed film.

FILM CHECKER
The individual who inspects film for defects and damage.

FILM-CLEANING MACHINE
A machine that cleans film by pulling it through an ultrasonic cleaning device or through a cloth saturated with solvent. (See also Ultrasonic Film Cleaner)

FILM CLIP
A short piece of film from a feature film, usually inserted into a telecast for promotional purposes.

FILM EDITOR
(Also Editor, Cutter)
The person responsible for arranging and assembling a film.

FILM FESTIVAL
The screening of many different films on successive days in one city, which culminates in awarding honors to the winning productions. One of the most well known is the Cannes Film Festival held annually in Cannes, France.

FILM GATE
The combination of pressure plate and aperture plate in a camera or projector.

FILM GAUGE
The size of standard motion picture film as measured in width and indicated by gauge in millimeters. The most common film gauges are 8mm, 16mm, 35mm, and 70mm.

FILM HANDLER'S GLOVES
The soft, lint-free cotton gloves worn during any handling of original film to prevent scratches and fingerprints, worn especially during certain stages of the editing process.

FILM LEADER
(1) A strip of blank film attached to film, used for equipment-threading purposes; (2) a strip of blank film attached to film, and used to create spaces during editing of the workprints and the preparation of A and B rolls. (See also Leader)

FILM LIBRARY
See Film Archive.

FILM LOADER (Also Loader)
The member of the camera crew whose chief responsibility is the loading of camera magazines.

FILM LOOP (Also Loop)
(1) A small amount of slack film between sprocket rollers and intermittents to keep the film from being torn; (2) a large amount of film in an endless roll used for continuous projection as in a cartridge; (3) large lengths of film or audio track spliced end to end and run continuously through a printer or audio playback. (See also Loop, Looping)

FILM MAGAZINE
A light-proof container that holds unexposed film and feeds it into a camera or processor without exposing it.

FILM NOIR
A film genre typified by characters of the underworld or with themes of violence and dark passion.

FILM PHONOGRAPH
An instrument that reproduces sound from an audio track for rerecording or playback.

FILM PICKUP
A process whereby motion pictures are electronically screened for image transmission on television equipment.

FILM PLANE
The front surface of film as it is held in the camera during exposure or at the gate of a projector.

FILM PROJECTOR
See Projector.

FILM RUNNING SPEED
The rate at which film runs through a camera or projector; expressed in meters per minute, feet per minute, or frames per second.

FILM SPEED
A film's sensitivity to light, or ASA. The lower the ASA, the slower or less sensitive the film. (See also Exposure Index, Fast, Forcing, Slow Film)

FILM STOCK
Raw film stock and the various sizes in which it is available.

FILM STORAGE
A holding room or vault, usually temperature- and humidity-controlled, where film is stored when not in use.

FILM STRUCTURE
The way in which time (present, past, and future) is presented in a film. (See also Diegesis, Dramatic Time, Expansion of Time, Filmic Time and Space)

FILM STUDIO
See Studio.

FILM TALENT AGENCY
An agency that seeks performers for projects and sends prospective performers to read for producers and casting directors in the hope of the actor getting a part. Agencies receive ten percent commission from the performer's salary for their services.

FILM TRANSPORT
The mechanisms made up of rollers and sprockets that move the film through the camera, processor, or printer from the supply reel to the take-up reel.

FILM TREATMENT
(1) A rough breakdown of a proposed presentation idea for a film; (2) the removal of excess moisture from processed film, followed by lubrication; (3) the repair of damaged film.

FILMIC TIME AND SPACE
The element of time as created by dramatic design, including expanding, condensing, or eliminating it for the screen; as well as shifting from past to future time through flashbacks and flashforwards. (See also Diegesis, Dramatic Time, Expansion of Time, Film Structure)

FILM VAULT
See Vault.

FILM WIDTH
See Gauge.

FILMOGRAPH
A film composed of still photos rather than live action.

FILMOGRAPHY
A list of all the films, dated and sequenced, in which one has appeared, directed, produced, or had a key function.

FILMSTRIP
A strip of film that contains images designed to be projected one frame at a time.

FILTER
(1) Any glass or gelatin optical attachment used in front of a luminaire, or with a lens system in a camera or printer to create effects, balance, absorb certain colors, or change color values; (2) an electronic device used to pass desired frequencies while rejecting others in sound equipment. An audio effect to give the voice the sound of being heard over a telephone or some faraway means of communication; (3) mechanical or chemical devices used to remove sludge from film processing solutions. (See also Absorption Filter, Behind-the-lens Filter, CC Filter, Contrast Filter, Conversion Filter, Daylight Conversion Filter, Diachroic Filter, Effect Filter, Fluorescent Light Filter, Fog Filter, Gel, Glass Filter, Graduated Filter, Haze Filter, Light-balancing Filter, Light Source Filter, Night Filter, Polarized Filter, Polaroid Filter, Sky Filter, Star Filter, Subtractive Process. Ultraviolet Filter, Viewing Filter.

FILTER FACTOR
A number that represents the amount of light absorbed by an optical filter and that must be compensated for before filming.

FINAL CUT
The original negative when it has been cut to conform with the workprint; the final edited film, ready to be dubbed and scored.

FINAL SHOOTING SCRIPT
The final draft of a script that has been approved for shooting.

FINAL TRIAL COMPOSITE
A print ready for release; a composite film incorporating all corrections and changes specified in previous composites.

FINDER
A viewfinder that gives the approximate field seen by the camera lens.

FINE CUT
A workprint almost completely edited and close to the stage of final trial composite.

FINE GRAIN
Film emulsion in which the silver particles are extremely small.

FINE-GRAIN DUPLICATE NEGATIVE
A duplicate negative made either by the reversal process from the original black-and-white negative, or from a master positive printed from the original black-and-white negative.

FINE-GRAIN MASTER POSITIVE

In black-and-white film processing, an intermediate step after the original negative, used for the printing of optical effects to protect the original.

FINGER

A narrow rectangular device used to cast shadows. (See also Dot, Flag, Gobo, Scrim, Target)

FIRE SHUTTER

A safety device at the gate of a film projector that prevents a burning film from spreading into the magazine.

FIRST CAMERA OPERATOR

The chief camera operator.

FIRST GENERATION DUPLICATE

See Generation.

FIRST GRIP (Also Key Grip)

A film production's chief stagehand.

FIRST MAN THROUGH THE DOOR

A colloquial expression meaning the villain of the plot.

FIRST PROPERTY MAN

A film production's chief prop person.

FIRST RUN

The first release of a feature film.

FIRST TRIAL PRINT (Also Answer Print, First Trial Composite Print, Trial Composite Print)

The first print of a film that is screened by the producer, director, and various other members of the production company in order to determine if the timing, color, and sound are acceptable.

FISHEYE LENS

An extreme wide-angle lens in which close images appear greatly distorted, and in which an entire landscape can be captured in the frame. (See also Lens)

FISHPOLE

A long, hand-held, portable pole upon which a mike is mounted. A fishpole is used in situations where a boom is not practical.

FITTING FEE

An additional sum paid to a performer for a wardrobe fitting.

FIX

(1) To prevent the developed image from fading by use of a chemical bath. A film with the images fixed, or stabilized, by the process of fixation; (2) to establish a fact for the viewer; (3) an unmoving camera during a shot.

FIXATION (Also Fixing)

The stabilizing process that "fixes" the image on the film base; a step that follows developing, in which the film is run through a chemical bath. Fixation terminates a film's sensitivity to light.

FIXED FOCUS

An adjustment made during manufacture to allow for permanent maximum depth of field. (See also Focus)

FIXED FOCUS VIEWFINDER
A viewfinder that allows for a sharply defined image regardless of the distance from the camera.

FIXER (Also Hypo)
The solution that stabilizes the developed image on a film.

FLAG
A device or small rectangular gobo made of plywood or cloth that is stretched on a frame, used to cast shadows. (See also Dot, Gobo, Finger, Target, Vignetting)

FLANGE
A circular device used on a rewinder upon which a plastic core is mounted for winding and handling film.

FLARE
Light streaks or other spots, haze, or fogging on film caused by light leaks in the magazine or camera or reflection of light from the surface of the lens or camera.

FLASH CUTTING
See Fast Cutting.

FLASH-FORWARD
(Also Flash Ahead)
A shot or scene depicting action that takes place in the future or will be seen later in the film.

FLASH FRAME
One frame unexpectedly inserted into a shot providing a different image for an instant.

FLASH PAN
See Swish.

FLASH POT
A ceramic pot used in special effects in which a flammable material or substance is electrically ignited to produce smoke or an explosion of light.

FLASHBACK
A scene that jumps back in time, out of chronological order of the script. Usually added to show a scene that has occurred earlier in the film or a scene that shows an event from the past; any shot that depicts action that took place before the film's present time.

FLASHING (Also Reexposure)
To expose film to a weak light either before or after camera exposure, but before processing, in order to reduce contrast; (2) reexposing reversal positive film after partial processing.

FLAT
(1) A section of set used as background; a wooden frame usually eight to ten feet high, and from six inches to twelve feet wide over which cloth, canvas, or other material is stretched and painted, wallpapered, or otherwise designed by the art director, as background (See also Working Drawings); (2) lifeless, lacking in emotion, i.e., a flat performance.

FLAT FEE
See Contract Player.

FLAT GLASS (Also Matte Glass)
Specially manufactured glass sheets that do not cause optical

aberration; do not reflect light from windows or pictures into the lens.

FLAT LIGHT
A shadowless, evenly cast frontal light that lacks contrasts and usually emanates from soft light sources.

FLATBED EDITING MACHINE (Also Flatbed Editor, Horizontal Editor)
A table with various devices used in editing film and soundtracks, i.e., sprocket drives, heads for picture projection and sound reading, and other features that are used in film editing.

FLICK (Also Flicks, Flicker)
A slang term for "the movies."

FLICK PAN
See Swish.

FLICKER
Visible flickering variations in the light beam being thrown on a screen by a projector.

FLIP CARDS
Slang expression for Cue Cards.

FLIP LENS
An attachment to an optical printer used to make flip-overs. (See also Lens)

FLIP-OVER (Also Flip-Over Wipe, Flip Wipe, Optical Flop, Turnaround)
An optical effect giving the viewer the impression that the picture has been turned over, either horizontally or vertically, revealing another picture on its back side.

FLIPPERS
See Barndoors.

FLOATING TRACK
See Wild Track.

FLOCK PAPER
Nonreflective, opaque paper used for mattes in special effects work.

FLOODLIGHT (Also Flood)
Any luminaire, or light, that casts a wide, scattered beam or flood of light, as distinguished from a spotlight that casts a specific pattern of light. (See also Striplight)

FLOODS
A short term for floodlights.

FLOOR MANAGER
The crew member who is the director's representative on the studio floor and is in constant two-way communication with the director. (See also Stage Manager)

FLOOR PLAN
(Also Ground Plan)
A scale drawing of a set as seen from above.

FLOP-OVER
See Flip-Over.

FLUB (Also Fluff)
To make a mistake while speaking lines of dialogue, for example, mispronouncing a word or forgetting lines.

FLUID HEAD
A type of tripod head that smooths

any jerky horizontal and vertical rotations by forcing a fluid through narrow channels that acts like a cushion.

FLUORESCENT LIGHT FILTERS

Filters used to improve color tones when using color film in fluorescent lighting conditions. (See also Filter)

FLUTING

A condition that results when film edges become bent due to swelling as a result of tight winding and high humidity.

FLUTTER

(1) An unwanted unsteadiness in the picture caused by movement of the film in the optical axis of the camera, printer, or projector; (2) a distortion in the sound caused by erratic motion of a component during recording or reproducing the sound.

FLUTTER SOUND

See Wow and Flutter.

FLUX

The measurement in lumens of light present.

FLY (Also Flying)

Hanging backdrops or scenery by cables in the fly loft above a stage; also a shortened term for fly loft.

FLY LOFT

The space directly above a stage.

FO

Abbreviation for Fade-Out.

FOCAL LENGTH

The distance between the main focus point of a lens surface and its optical center, or point of focus, when the lens is focused at infinity; the distance of the focus from the surface of the lens. The greater the focal length of a lens, the farther it can "see" clearly, the greater its telescopic capability. (See also Short Focal Length Lens)

FOCAL PLANE

The plane in which an image is formed clearly.

FOCAL POINT

The point at which the action field is clearly focused when the lens is focused at infinity; the intersecting point between the optical axis of the lens and the focal plane.

FOCUS

A term that in general industry use refers to the focusing point at which the image has reached maximum sharpness, definition, and clarity; a clear picture. (See also Critical Focus, Defocus, Deep Focus, Depth of Field, Differential Focus, Fixed Focus, Follow Focus, Pan Focus, Plane of Critical Focus, Pull Focus, Rack Focus, Reflex Focusing, Selective Focus, Shallow Focus, Split Focus, Universal Focus)

FOCUS PULLER

(1) A camera operator's assistant who keeps the lens in focus during shots; (2) a handle attached to the lens focusing ring that facilitates follow focus.

FOCUSING

Bringing the image into sharpness and clarity by rotating a focusing ring on a camera lens that adjusts the distance from the lens to the film.

FOCUSING VIEWFINDER
A viewfinder that, because of its shallow depth of field, must be refocused when the action's distance from the camera is altered.

FOG
(1) An obscuring of the visual image as a result of unwanted light exposure during camera loading or a leak within the camera itself; (2) the amount of developing possible when a film is processed even without exposure to light. (See also Fog Level)

FOG DENSITY
A characteristic density a film has when developed with no exposure to light, but caused by factors such as age and temperature.

FOG EFFECT
See Rumble Pot.

FOG FILTER
A lens filter that creates a soft, foglike effect by softening the image with a veiled, diffused atmosphere. (See also Filter)

FOG LEVEL
The amount of development possible or development capacity of a roll of film, even before exposure to light. Fog level rises as unexposed film ages.

FOGGING
(1) Light—A dark blur or film density created by unwanted light exposure; (2) chemical—an obscured visual image or film density created by the misuse of certain chemicals or excessive exposure to air during the development process.

FOGMAKER
A special effects device that creates artificial fog.

FOLLOW F-STOP
When moving from a strongly lighted area to a dark one, the exposure is changed by moving from one f-stop to another, to compensate for the change in brightness; to change exposure during a shot.

FOLLOW FOCUS
(Also Rack Focus)
Keeping the shot in focus while the distance between the actor and the camera changes by continuously refocusing the lens during the shot; changing the focus during a shot to maintain sharpest clarity of the image. (See also Critical Focus, Focus)

FOLLOW SHOT
A moving shot, accomplished by dollying or trucking; a shot that follows the action. (See also Shot)

FOLLOW SPOT
A narrow-beamed spotlight aimed at moving performers by a follow-spot operator.

FOLLOW-SPOT OPERATOR
A light technician whose job it is to move a follow spot as a performer moves.

FOOT (Also Tail)
The end of a reel of tape or film.

FOOT-LAMBERT

The international unit of brightness that is one lumen per square foot of light from a radiating surface; or the uniform brightness of a perfectly diffusing surface from which light is emitted or reflected at the rate of one lumen per square foot.

FOOTAGE

(1) A colloquial term for film that has been shot and processed, or sound that has been recorded; (2) a term of measurement of amounts of film, whether exposed or raw stock.

FOOTAGE COUNTER

Any device used to measure the amount of film used by automatically counting the number of feet passed through equipment, especially a camera, but also printers and editing or projection units.

FOOTAGE-TIME CALCULATOR

A device that gives the scale or correlation between the length of film and its running time.

FOOTCANDLE

An international unit of measurement equal to the amount of light cast by one international candle on a square foot of surface, every part of which is one foot away from the light source.

FOOTCANDLE METER

A meter that measures light intensity in footcandles.

FORCING (Also Pushing)

To process film longer, which has the effect of increasing the film's speed to compensate for underexposure. (See also Film Speed)

FOREGROUND (FG)

The area in the action field closest to the audience.

FOREGROUND TREATMENT

An artistic arrangement of props or objects in the foreground of a shot to create a pleasant frame for the subject or action.

FOREIGN RELEASE

The major distribution of a feature film in countries outside the United States.

FORELENGTHENING

An effect that gives the appearance of an extension of depth in shots made with wide-angle lenses. It disappears if the image is viewed through the lens from its center of perspective.

FORESHORTENING

The exaggeration or contraction of depth that results from using a telephoto lens or when a subject gets too close to the camera lens. It disappears if the image is viewed from its center of perspective.

FORMAT

(1) The structure of a show, the way it opens, proceeds, and closes; (2) the general style of a script; the way in which the script is organized, i.e., number of scenes and length of script; (3) the size of a film stock and its perforations, along with the shape and size of the image

frame; (4) the size of video or audio tape, i.e., a 2-inch format indicates a show is to be taped on 2-inch tape, or that a system only plays 2-inch tape.

FOUNDATION LIGHT
See Base Light.

FOUR PLATE
See Two Plate.

FOUR-WALLING
A film distribution agreement in which a distributor pays an agreed-upon figure, usually the theater breakeven figure plus a profit, to an exhibitor. The distributor controls the ticket pricing and advertising for the film. All profits above the originally agreed payment belong to the distributor, in some cases even including concessions.

FOUR-WALL SET
A completely enclosed set in which the action and camera coverage take place within four walls. (See also Set)

FRAME
(1) An individual picture on a strip of film; (2) to physically arrange the performers in a scene; to compose a picture; (3) the area outlined by the viewfinder.

FRAME COUNTER
A device, included in a film counter, that counts the number of frames of film that passes through it. In 16mm there will be 40 frames per foot, in 35mm there will be 16 frames per foot.

FRAME GRABBER
See Frame Stopping Terminal.

FRAME LINE
(1) The line that separates each frame on a length of film; (2) the area outlined and framed by a camera's viewfinder.

FRAME RATE (Also Camera Speed, Projector Speed)
Usually expressed in FPS (frames per second), the rate or speed at which film passes the lens of a camera or projector.

FRAME STOPPING TERMINAL (Also Frame Grabber, Single Frame Terminal)
A device that enables a TV set to allow viewing of a still picture or single frame from a motion picture.

FRAMER
A mechanism that enables the operator to eliminate frame lines on the screen by raising or lowering them.

FRAMES PER FOOT (FPF)
The number of frames per foot of film; 35mm has 16 FPF, 16mm has 40.

FRAMES PER MINUTE (FPM)
The number of frames that in one minute pass through a projector or printer, or are exposed in a camera.

FRAMES PER SECOND (FPS)
The rate or speed at which film moves through a camera or projector. For example, 35mm film usually travels through a camera or projector at 24 frames per second, or 90 feet per minute.

FRAMING
The positioning and movement of a camera by the operator whereby superfluous action is eliminated, and the shot is framed with artistic composition. (See also Vignetting)

FRANCHISE
A contractual agreement between a cable television system and a city or county, defining the conditions of the installation and operation of the cable system.

FREE-LANCE PRODUCER
A producer not contracted exclusively to a major studio or production company, but who contracts independently on a production-by-production basis.

FREEZE FRAME
(1) Stopping a film or video in a projector or playback unit on one frame; (2) an optical effect of printing one frame several consecutive times, which appears to arrest the action and allows the image to be viewed without movement for as long as desired.

FREEZE FRAME PRINTING
A printing technique that requires special equipment, whereby a particular frame is printed several consecutive times, which stops the action by showing the selected still frame on the screen for as long as the repeat printing of the single frame continues.

FREQUENCY
The number of cycles per second. One cycle per second is called a Hertz (Hz).

FREQUENCY RESPONSE
The total range of sound, that is, the highest and lowest sounds that can be reproduced or recorded.

FRESNEL LENS
A lens used on spotlights to condense the light with a system of concentric prismatic ridges on the convex side, which greatly reduces bulk and weight. Lamps using this lens are commonly referred to as Fresnels. (See also Step-Prism)

FRICTION HEAD
A type of tripod head or camera mount that allows smooth movement of the camera for pans and tilts.

FRINGE TIME
One of the designated television time periods for the purpose of selling advertising and gathering ratings data. For Eastern, Mountain, and Pacific Time Zones, the slot is from 7:30 P.M. until 8:00 P.M., Central Time Zone fringe time is from 6:30 P.M. until 7:00 P.M. In all time zones, it is the half hour that just precedes prime time. (See also Daytime, Early Fringe Time, Family Hour, Late Fringe, Prime Access, Prime Time)

FRINGING
A loss of clarity around a matted background image that also usually results in a breakdown of colors.

FROM THE TOP
The director's order to begin at the beginning of the scene. In the early days of film and television, each scene began at the top of a new

page. That practice, however, is not followed in the industry today.

FRONT END
See Above-the-Line Costs.

FRONT LIGHTING
Lighting that originates from near the camera.

FRONT PROJECTION PROCESS
Any of several ways to project background images on a screen behind the actors and the action. Any image projected onto the actors is eliminated by supplemental lighting.

FROST
Translucent, solid, white, flat material that is used to diffuse light, also called Bridham's No. 1 after a commonly used name brand.

FULL APERTURE
The largest opening of a lens; the smallest f-stop number.

FULL COAT (Also Full Coat Magnetic Film)
Magnetic film that is completely covered on one side with an iron oxide coating, as distinguished from magnetic sound tape. (See also Magnetic Film)

FULL IMMERSION CONTACT PRINTER
A printer that completely immerses both the original film and the raw stock on which it is being printed in liquid at the time of contact to keep scratches from being printed. After printing the liquid is removed with squeegees and a dryer.

FULL SHOT
(Also Medium Long-Shot)
A shot that is framed just above the head and just below the feet of the subject. (See also Shot)

FUNGUS SPOTS
Blotches on the film emulsion caused by fungus.

FUZZ OFF
A softening of matte lines that reduces their visibility.

FUZZY
An image that is out of focus.

FX
Abbreviation used for "effects," as in sound effects (SFX).

G

GAFFER
The chief electrician on a film crew who is responsible for the lighting circuits and equipment.

GAFFER'S TAPE
A wide, strong adhesive tape used on the set to secure cables, stands, etc., and for a variety of other production uses.

GAG
A joke. (See also Pratfall, Sight Gag, Stunt)

GAIN CONTROL
The volume control on a sound unit or console.

GALVANOMETER RECORDER
An optical sound recorder with an attached galvanometer that is capable of changing light intensity and/or patterns of light beams as they strike the sound track.

GAMMA
A degree of contrast, measuring a negative's density and the extent of development; a lab reference term.

GANG SYNCHRONIZER
A device used to set up originals, workprints, and soundtracks for synchronization; consisting of more than one sprocket wheel and referred to as 2-gang, 3-gang, 4-gang, etc., depending on its capacity.

GANGSTER FILM
A film genre whose plot and characters revolve around the criminal underworld.

GARBAGE MATTE
A matte that not only includes the subject but also any additional equipment that was present when the matte was made. These will later be removed from the image in the final stages of the optical work. (See also Matte)

GATE
The fixed aperture plate and the moveable, spring-loaded pressure plate that aligns the film with the lens.

GATOR GRIPS
An alligator clamp used to anchor lightweight lamps to furniture, pipes, or pieces of equipment.

GAUGE
The term used for standard film width.

GEARED HEAD
A camera mount operated by a crank-and-gear mechanism that stabilizes the camera movement, especially in tilts and pans.

GEL (Also Gelatin, Jellies)
Transparent gelatin filters in sheets of various colors which, when placed in front of a clear or "white" light, changes the light to the color of the gel. (See also Bastard Amber, Filter)

GENEALOGY
The numerical history of duplicates made from the master (film or videotape); the first copy made from the original being "first generation." (See also Generation)

GENERAL CONTINUITY LINK
The basic theme of the story around which the show is built.

GENERAL RELEASE
The distribution of a film to theaters in all cities of the country.

GENERATION
A designation that defines how close a copy is to the original film or the master tape from which it derives; the first generation is the copies made from the original, the second generation is the copies made from the first generation, and so on.

GENEVA MOVEMENT
(Also Geneva Cross, Maltese Cross Movement)
A system in 35mm equipment in which a motion-transmitting mechanism holds successive frames of film at a set exposure or projection aperture in cameras, projectors, or step printers. This is usually a 4-channel system with the key elements set in a crosslike pattern, hence the term.

GENRE
A category of film in which the subject matter and, to some extent, the treatment of the subject matter, are predictably similar to other films of its type; western films, mystery films, horror films, and science fiction films are all examples of film genres.

GEOMETRY FILM
The emulsion position of the print as it relates to the left-right images on the picture; geometry changes in subsequent printings.

GESTURE
(1) A specific movement by a performer; (2) a dominant image or movement that provides the editor with a cue or basis on which to splice.

"GET IN CHARACTER"
A direction, sometimes used by the director for alerting actors when performance is about to begin.

GHOST
See Halation.

GHOST IMAGE
A double outlining or multiple image seen on a television screen, caused by two or more signals arriving at slightly different times. Delays can be caused by signal reflections from tall buildings. (See also Halation)

GIMBAL MOUNT
(Also Gimbal Tripod)

A camera support consisting of a free-swinging mount from which a pendulum weight has been suspended. This type of camera mount is used on tilting surfaces as it keeps the camera level by compensating for the pitch and roll.

GIMMICK
See Plot Gimmick.

GLASS FILTER
A filter consisting of a gelatin sheet between two sheets of glass for color or neutral filter effect; also, a solid piece of colored glass. (See also Filter)

GLASS PAINTING
A painting on glass, generally background, used to make a glass shot.

GLASS SHOT
A shot taken through a pane of glass upon which has been painted background that is too expensive to build, titles, or artwork. (See also Grease-Glass Technique, Shot)

GLOVING SOUND
See Velveting Sound.

GLOW LIGHT
A dim light projected along the edge of the subject, e.g., the glow of a candle.

"GO TO BLACK"
A television term usually commanded by the director, meaning that the screen is to become totally black; a total fade-out; the equivalent of the theater's "curtain."

GO TO TABLE
An informal term that refers to the first time a script is read by all involved in this stage of production; no blocking is done and all are usually gathered around a table simply to read the script aloud.

GOBO
Black wooden boards or cloth screens (a large flag) or any opaque material used in blocking direct light from the camera lens, thus eliminating the possibility of flare. (See also Dot, Finger, Flag, Scrim, Target)

GOBO STAND (Also Century Stand, C Stand)
A stand made of metal used to position various grip equipment and lighting accessories.

GOFER (Also Gopher)
Literally, an employee who goes for this or that; a courier or assistant who handles nontechnical/noncreative tasks and errands.

"GOING OFF"
A directive term used in a script or by command indicating to the actor that the line should be delivered as he is moving away from the microphone.

GOLDWYNISM
Hollywood producer Samuel Goldwyn was well known for his many malapropisms, which became so well known that the term Goldwynism was coined to describe them. Two of the more well known are: "A verbal contract isn't worth

the paper it's printed on," and "Include me out."

GOPHER
See Gofer.

GOTHAM
A term used to refer to New York City.

GRADATION
The degree of change in tones in a photograph.

GRADER
The lab technician in charge of the grading or timing process. (See also Grading)

GRADING (Also Timing)
The first step before printing a film; at this stage the correct printer lights, exposure settings, and color corrections are determined to improve the quality of the original footage.

GRADUATED FILTER
A supplementary lens or filter with color in one portion, which gradually bleeds out to clear. (See also Filter)

GRAIN
The visible particles of silver that have clumped together in processing and have become embedded in the gelatin of the film's emulsion, thus appearing in the picture as tiny grains.

GRAININESS
The amount of visible silver particles that causes the texture of the film's picture to appear granular, as in the surface of a piece of sandpaper. The tiny particles are formed when silver grains clump together and embed in the gelatin of the film's emulsion; the more visible they become, the more graininess the picture has.

GRANDFATHERING
Allowing a television cable system to be exempt temporarily from federal rules about franchising requirements. To qualify, the system must have been established before March 31, 1972, must currently be operating and must serve at least fifty subscribers. If the system is operating under the grandfathering clause, it must comply with all current federal rules within five years from the date of notice or at the expiration of its current franchise, whichever occurs first. (See also Cable)

GRAPHICS
All printed or hand-lettered materials such as titles, charts, and graphs; usually handled by the art department. (See also Artwork, Videographics)

GRAY BASE FILM
(Also Gray Base Stock)
Film stock whose base is gray, thereby reducing halation, as opposed to film using rem-jet backing for antihalation. If rem-jet backing is used, someone must actually buff it off during processing, a step that can be omitted if gray base is used.

GRAY CARD
(Also Standard Gray Card)
A matte-finish card, gray on one side, white on the reverse, used to obtain light meter readings. The gray side with 18 percent reflec-

tance is read with a reflected light meter, while the white side is read when the light is too dim to allow a reading from the gray side.

GRAY SCALE
A test chart of various shades of gray from white to black; generally used in exposure tests.

GREASE-GLASS TECHNIQUE
A technique whereby a shot is made through a glass that has been smeared with a greasy substance (most often Vaseline or KY jelly) to blur part of the action field. (See also Glass Shot, Shot)

GREASE PAINT
(1) A slang term for all makeup used by actors; (2) a grease-based makeup especially in tones other than normal skin shades, such as clown makeup in white, red, and blue. (See also Makeup)

GREEN PRINT
A print that has been improperly processed and dried. Usually, because of greater than normal surface friction, a green print will be damaged during projection; also, a just-processed, still-wet print.

GREEN ROOM
A preparatory room where performers wait just before going on camera.

GREENERY
Decorative shrubbery, flowers, or plants, whether real or artificial, that are used on the set or location during shooting.

GRID (Also Lighting Grid, Pipe Grid)
Interlocking steel pipes that are hung horizontally from the ceiling and are used to support lights and scenery.

GRIPS
The crew members responsible for varied jobs around the set, such as laying camera tracks, building scaffolding, handling gobos, general set constructions; the carpenters and general handymen on a set.

GRIP CHAIN
A lightweight chain used by the grips to hang scenery, secure certain props, and for many other miscellaneous uses on a set.

GRIP TRUCK
A small, nonmotorized push truck that is used to carry lighting equipment on a set.

GROUND GLASS
The focusing screen in a camera, made of frosted translucent glass, upon which the image before the camera appears.

GROUND-GLASS VIEWFINDER
A viewfinder in which a ground-glass screen replaces an optical system to display the action. (See also Viewfinder)

GROUND PLAN
See Floor Plan.

GROUP SHOT (Also Three Shot)
A relatively close shot of three or more persons. (See also Shot)

GUIDE TRACK

A low-quality sound track recorded with the picture only as a guide for postsynchronization. (See also Scratch Track)

GUILDS

A group of people who perform similar tasks or have similar jobs in the industry and who have banded together for their protection, camaraderie, or united interests. Following is a list of the major guilds in Los Angeles and New York.

In Los Angeles:

Academy of Motion Picture Arts & Sciences
8949 Wilshire Blvd.
Beverly Hills, CA 90211

Academy of Television Arts & Sciences
4605 Lankershim Blvd., Suite 800
No. Hollywood, CA 91602

AFTRA
1717 N. Highland Ave.
Hollywood, CA 90028

American Cinema Editors, Inc.
422 S. Western Ave.
Los Angeles, CA 90029

American Film Institute
2021 N. Western Ave.
Los Angeles, CA 90027

American Guild of Variety Artists (AGVA)
6430 Sunset Blvd., Suite 503
Hollywood, CA 90028

American Society of Cinematographers
1782 N. Orange Drive
Hollywood, CA 90028

Association of Film Craftsmen
NABET, Local 531, AFL-CIO
1800 N. Argyle St., Suite 501
Hollywood, CA 90028

Association of Talent Agencies
9255 Sunset Blvd.
Suite 930
Los Angeles, CA 90069

Black Stuntmen's Association
P.O. Box 1773
Hollywood, CA 90028

Broadcast TV Recording Engineers
Local 45, IBEW
3518 Cahuenga Blvd., W.
Hollywood, CA 90068

Choreographers Guild
256 S. Robertson Blvd.
Beverly Hills, CA 90211

Costume Designers Guild
Local 892, IATSE
11286 Westminster Ave.
Los Angeles, CA 90066

Directors Guild of America
7950 Sunset Blvd.
Los Angeles, CA 90046

Film Advisory Board
1727 N. Sycamore Ave.
Hollywood, CA 90028

Film Technicians
Local 683, IATSE, AFL-CIO
6721 Melrose Ave.
Los Angeles, CA 90038

GUILDS

Hollywood Stuntmen and
 Production Workers Union
6311 Romaine Street
Los Angeles, CA 90038

IATSE & MPMO
7715 Sunset Blvd., Suite 210
Los Angeles, CA 90046

International Sound Technicians
Local 695, IATSE, MPMO,
 AFL-CIO
11331 Ventura Blvd.
Studio City, CA 91604

Makeup Artists & Hair Stylists
Local 706, IATSE, MPMO
11519 Chandler Blvd.
No. Hollywood, CA 91601

Motion Picture Association of
 America, Inc.
8480 Beverly Blvd.
Los Angeles, CA 90048

Motion Picture Costumers
Local 705, IATSE, MPMO,
 AFL-CIO
1427 N. La Brea Ave.
Hollywood, CA 90028

Motion Picture Editors Guild
Local 776, IATSE
7715 Sunset Blvd., Suite 100
Los Angeles, CA 90046

Motion Picture Illustrators &
 Matte Artists
Local 790, IATSE
7715 Sunset Blvd., Suite 210
Los Angeles, CA 90046

Motion Picture Screen
 Cartoonists
Local 839, IATSE
4929 Lankershim Blvd.
No. Hollywood, CA 91602

Motion Picture Studio
 Cinetechnicians
Local 789, IATSE, AFL-CIO
1635 Vista Del Mar Ave.
Hollywood, CA 90028

Motion Picture Studio Grips
Local 80, IATSE, AFL-CIO
6926 Melrose Ave.
Los Angeles, CA 90038

Producers Guild of America,
 Inc.
292 S. Lacienega Blvd.
Beverly Hills, CA 90211

Scenic & Title Artists
Local 816, IATSE
7429 Sunset Blvd.
Los Angeles, CA 90046

Screen Actors Guild (SAG)
7750 Sunset Blvd.
Los Angeles, CA 90046

Screen Cartoonists Guild
Teamsters Local 986
1616 W. 9th St.
Los Angeles, CA 90015

Screen Extras Guild, Inc. (SEG)
3629 Cahuenga Blvd. West
Los Angeles, CA 90068

Script Supervisors
Local 871, IATSE
7715 Sunset Blvd., Suite 208
Los Angeles, CA 90046

GUILDS

Set Designers & Model Makers
Local 847, IATSE
7715 Sunset Blvd., Suite 210
Los Angeles, CA 90046

Society of Motion Picture and
   TV Art Directors
Local 876, IATSE
7715 Sunset Blvd.
Los Angeles, CA 90046

Story Analysts Guild
Local 854, IATSE
7715 Sunset Blvd., Suite 210
Los Angeles, CA 90046

Stuntmen's Association
4810 Whitsett Ave.
N. Hollywood, CA 91607

Stunts Unlimited
3518 Cahuenga Blvd. West
Los Angeles, CA 90068

Stunt Women of America
202 Vance Street
Pacific Palisades, CA 90272

Theatrical Stage Employees
Local 33, IATSE, AFL-CIO
4610 Lankershim Blvd.,
   Suite 833
No. Hollywood, CA 91602

Writers Guild of America—West
8955 Beverly Blvd.
Los Angeles, CA 90048

In New York:
   American Federation of TV-
      Radio Artists
   1350 Avenue of the Americas
   New York, NY 10019

American Guild of Variety
   Artists
1540 Broadway
New York, NY 10036

Cartoonists Guild
156 West 72nd Street
New York, NY 10023

Directors Guild of America, Inc.
110 West 57th St.
New York, NY 10019

International Alliance of
   Theatrical Stage Employees &
   Moving Picture Operators
   U.S. and Canada
1515 Broadway
New York, NY 10036

Makeup Artists & Hair Stylists
Local 798
1790 Broadway
New York, NY 10019

Motion Picture Association of
   America, Inc.
522 Fifth Ave.
New York, NY 10036

Motion Picture Film Editors
Local 771, IATSE
630 Ninth Ave.
New York, NY 10036

Motion Picture Laboratory
   Technicians
Local 702, IATSE
165 West 46th Street
New York, NY 10036

Motion Picture Studio
   Mechanics
Local 52, IATSE
221 West 57th Street
New York, NY 10019

GUILDS

National Academy of Television
  Arts and Sciences
110 West 57th Street
New York, NY 10019

National Association of
  Broadcast Employees &
  Technicians
111 West 50th Street
New York, NY 10020

Radio & Television Broadcast
  Engineers Union
Local 1212, IBEW
230 West 41st Street
New York, NY 10036

Screen Actors Guild
551 Fifth Ave.
New York, NY 10017

Screen Cartoonists
Local 841
25 West 43rd St.
New York, NY 10036

Studio Mechanics
Local 52
221 West 57th St.
New York, NY 10019

Writers Guild of America-East
222 West 48th St.
New York, NY 10036

H

HAIR LIGHT
See Kicker Light.

HAIR STYLIST
The production staff person who styles and maintains the performers' hair, hairpieces, or wigs during the production.

HALATION (Also Ghost)
A usually unwanted halolike illumination outlining images on film or tape, caused either by light that shines too brightly into the lens or by exposure light reflecting from the film base into the emulsion; the latter is easily controlled by using film with an antihalation coating or dye. (See also Antihalation Backing, Ghost Image, Gray Base Film)

HALF SHOT
See Medium Shot.

HALOGEN LAMP (Also Quartz Lamp, Quartz-Iodine Lamp, Tungsten-halogen Lamp)
A tungsten bulb that has a longer life and maintains color temperature better than a nonhalogen lamp. The tungsten in the bulb is evaporated from the filament then recycled and returned to the filament by a halogen element such as iodine or chlorine.

HAND CAMERA
A small production camera that can be held in the hand during operation.

HAND CRANK
The crank used to activate and power a hand-cranked camera. (See also Hand-cranked Camera)

HAND-CRANKED CAMERA
A manually operated camera that is powered by turning a crank at a steady pace. This camera was primarily used for the filming of silent pictures.

HAND HELD
A shot filmed by a camera operator hand-holding a camera; also the amateurish wavering of the action on the screen that results from unsteady holding of a camera.

HAND PROPERTIES
(Also Hand Props)
Props that can be carried readily or used in the hands of the actors during filming, such as a doll, apple, flower, pen, or newspaper.

HANDLE
(1) A grasp of the situation; (2) an element of plot that explains a certain situation or twist in the story; i.e., the fact the character dropped out of medical school just before graduation is a handle that would enable the audience to believe the

character could perform emergency surgery.

HANDLEBAR MOUNT
A camera mount consisting of two handles located near the front that allows the camera operator to use both hands to control the camera movement.

HARD FLUFF
Informational entertainment.

HARD-FRONT CAMERA
A camera with one hole for the attachment of one lens, as opposed to a moveable turret that allows for the use of several lenses.

HARD LIGHT
Light that is produced in a narrow beam, thus creating sharply defined shadows. Hard light doesn't scatter, unlike soft light, which diffuses readily; hard light is used to approximate sunlight on a set. (See also Soft Light)

HARD-TICKET ROADSHOW
(Also Hard-Ticket Show or Attraction)
A production for which each seat is numbered and reserved; usually a premiere or first-run feature film exhibition for which seats are assigned by printing the seat number on the ticket.

HARDENER
A chemical used in processing that causes the gelatin holding a photographic emulsion to harden.

HARDWARE
The equipment used to produce and broadcast software, including cameras, videotape recorders, monitors, and all equipment related to production, storage, distribution, or reception of electronic signals. (See also Software)

"HAVING HAD"
A phrase used when informing a performer of his/her call, in order to let that performer know that he/she should eat before reporting to the set. For example, if a performer is to report at 1:00 P.M., he/she will be told to "report at 1:00 P.M., having had..."

HAYS OFFICE
The forerunner to the Motion Picture Association of America, the Hays Office was the popular term for the Production Code Administration that regulated what was morally acceptable in a film during the 1930s, '40s, and '50s. Often representatives would be on the set and, when a conflict between them and the director/producer occurred, the production was shut down until the situation was rectified. Mae West was particularly scrutinized; however, all pictures were affected by the Code. It dictated that twin beds be used in a scene even if a couple was married. This carried over into television, which is why twin beds were used in "I Love Lucy" in the Ricardo bedroom. In love scenes, the gentleman had to keep one foot on the floor at all times. The Hays Office closed in 1966.

HAZE FILTER (Also Haze-Coating Filter, Ultraviolet Filter)
A specially designed filter that

absorbs ultraviolet light, making it a necessity when filming in certain outdoor sunlight conditions. (See also Filter)

HAZELTINE
A machine that projects a film negative as a positive image on a screen during the grading, or color balancing, process allowing for easier refinement of the colors as specified by the filmmaker.

HEAVY
The villainous character or antagonist of a story's plot.

HELICOPTER MOUNT
See Copter Mount

HERTZ
The unit of audio frequency. One hertz equals one cycle per second. 1,000 Hz = 1 KHz (kiloHertz); 1,000,000 Hz = 1 MHz (mega Hertz). (See also Frequency)

HIATUS
A scheduled interruption or vacation in production, especially in television production schedules of more than thirteen weeks; every few weeks a week off is usually scheduled when budget and time commitment permit.

HIGH-ANGLE SHOT
A shot made by placing the camera above the subject with the camera looking downward toward the action; a downward angle shot. (See also Camera Angle, Shot)

HIGH CONTRAST
A sharp gradation of black and white tones on a photographic image with few intermediate gray tones; a harsher, clearer, black and white. (See also Low Contrast)

HIGH-INTENSITY ARC
(Also High-Intensity Carbon Arc; High Arc Light)
A very bright light produced by a luminaire that uses a transformer and carbon arc for intensity.

HIGH KEY (High Key Lighting)
A lighting style using few dark tones. The opposite of low key lighting; a bright set in which the sets and lighting effects are designed to lower contrast, often assisted by light-colored costumes and backgrounds. In this style of lighting, the majority of the scene is in highlights and the fill lights are almost as bright as the key light. (See also Low Key, Light)

HIGH LEVEL
A term that refers to volume, tones, or contrast that register in the upper scale of measurement; if audio, it is loud sound.

HIGH-PASS FILTER
An audio circuit electronic filter that weakens or eliminates all frequencies below a selected level; frequencies above that level are picked up with clarity.

HIGH-SPEED CAMERA
A specially designed camera used to achieve a slow-motion image on the screen. High speed cameras film at speeds anywhere from approximately 3,000 frames per second to 16,000 frames per second. Twenty-four frames per second is normal. When the film is projected at nor-

mal speed, the action appears to be extremely slow. This type of camera can be used for special effects or frame-by-frame analysis of the action, but generally it is used for scientific purposes, such as photographing a rocket launch. (See also Slow Motion)

HIGH-SPEED CINEMATOGRAPHY
(Also High Speed)

Filming done with a high speed camera to achieve slow motion photography when the film is projected at normal speed. Generally this is useful for obtaining scientific data. It can also be very effective for creating certain dramatic effects. (See also Slow Motion)

HIGH-SPEED FILM

(1) Film that is manufactured with extra perforations for use in high speed cameras to achieve a slow motion effect; (2) a film that has been shot with high-speed cinematography for normal projection and thus appears to slow down the action; (3) extremely light-sensitive film. (See also Slow Motion)

HIGHLIGHT DENSITY

The indication or measurement of the lightest part of a positive or the darkest part of a negative.

HIGHLIGHTS

The brightest parts of a filmed image; they have the heaviest density on the negative and are the lightest parts of the positive image.

"HIT 'EM ALL" (Also "Hit the Juice")

A directive to switch on all the lights to be used during that scene; to turn on immediately the electricity that is used for lights and/or any electronic effect being used in the scene.

"HIT THE MARK"

The directive given to the performers to move to a previously established mark because of lighting and camera focus that was adjusted earlier; often the mark is a piece of gaffer's tape stuck to the floor.

HMI LAMP

A ballast-controlled lamp capable of producing high-quality light that matches sunlight.

"HOLD"

(1) A directive to reserve that take for possible use in the finished production; (2) In a script, a scene or shot that is held in reserve for possible use at the time of shooting; (3) a directive to stop using temporarily the effect or prop specified in the command.

HOLD CEL

An animation cel that is filmed as is for many frames without change, used with cels that change frequently, thus eliminating the need to redraw elements that remain the same from frame to frame. (See also Animation)

HOLLYWOOD BLACKLIST

Created as a result of the 1947 investigations of the House Committee on Un-American Activities, the existence of this list is still denied by many. It was said to contain either the names of those "acceptable" for hire or of those persons whom the film industry should not hire because of possible un-American activities in their private lives.

HOLLYWOOD TEN
The ten film industry men who refused to testify before the House Un-American Activities Committee in 1948 and as a result served prison terms. The ten were: Alvah Bessie, Herbert Biberman, Lester Cole, Edward Dymtryk, Ring Lardner, Jr., John Howard Lawson, Albert Maltz, Samuel Ornitz, Adrian Scott, and Dalton Trumbo.

HOLOGRAM
A lifelike, three-dimensional image that can be projected in midair without a screen, produced through a combination of lasers and special photography.

HOLOGRAPHY
Special laser photography used to make holograms.

HONEYWAGON
The slang term for portable toilet trailers used on location.

HOOK
A plot gimmick or twist designated to hook, or maintain the interest of the audience. (See also Plot Gimmick)

HORIZONTAL EDITOR
See Flatbed Editing Machine.

HORROR FILM
A type of film in which the hero or heroine is faced with danger from monsters, the supernatural, or from bizarre events.

HORSE
A film dispenser; a simple editing-room device that dispenses film. Most often used to supply leader.

HORSE OPERA
An informal, slang term for a Western film, usually refers to a "B" Western. (See also Western)

HOT FRAME
An overexposed frame that usually occurs when the shutter is open longer than necessary at the beginning or end of a shot.

HOT
(1) Too brightly lit for shooting (see also Hot Spot); (2) a set ready for shooting (see also Hot Set); (3) a performer or effect that is much in demand.

HOT PRESS TITLES
Titles made from heated type pressed upon transfer paper, which is then pressed onto a title board or cel; the letters can be black, white, or colored.

HOT SET
A completely finished set ready for shooting in every detail including placed props. Those in the production, other than performers who are cued to do so, are not to walk into, sit in, or touch the set for any reason.

HOT SPLICE
(1) To splice with a hot splicer; (2) the splice made by a hot splicer. (See also Hot Splicer)

HOT SPLICER
A film splicing machine that uses heat to hurry the drying action of the film cement.

HOT SPOT
An area in a scene that is too bright from excessive light or reflection; an exposure or area of light that must be toned down.

HOUSE LIGHTS
The lights in the audience area of a soundstage, auditorium, or area of production where a live audience can be present for a production; in general, all lights in an area of production other than those specifically used on the set.

HOW-TO-DO-IT FILM
An informal term that refers to an instructional film, showing a skill or process, step by step.

HUB
A cable television term that refers to one of at least two headends that are interconnected to provide cable service over a wide area. (See also Cable, Headend)

HUE
A formal term for color and its gradations.

HUT
An abbreviation used by the television rating's services to indicate Homes Using Television, or the homes in which television is being watched at any given time.

HYPE
(1) Publicity gimmicks, advertising actors' personal appearances, or any created method of promoting or pumping the sales appeal of a production; (2) a slang reference to inflated, nonfactual information used to create a false impression; i.e., "the audience reaction was hyped for the press" or "the agent's praise of the star's talent is mainly hype to sell the producer."

HYPERFOCAL DISTANCE
The focus distance at which a lens is capable of giving the greatest depth of field; a combination of critical focus and f-stop that creates a sharp image from half the distance of critical focus to infinity. (See also Critical Focus)

HYPHENATE (Hyphen)
A person who performs more than one major production function; i.e., producer-director, producer-writer, writer-director, writer-director-producer. Such a person is called a hyphenate.

HYPO (Also Fixer)
A chemical used in the developing process that (1) stops the development; (2) removes undeveloped silver; and (3) hardens the gelatin of the emulsion.

Hz
An abbreviation for Hertz.

IATSE

The commonly used abbreviation for the International Alliance of Theatrical and Stage Employees, one of the most powerful unions involved in major film or tape production, comprised of craftsmen and technicians from key areas of the crew, including stagehands, makeup artists, and wardrobe handlers.

ICONOSCOPE

An electron tube that changes optical images into electric impulses for the purpose of television signals; an early camera picture tube, replaced by the image orthicon tube.

ID

Identification of a station or network's call letters by announcing them on-the-air, as required by the FCC.

IDIOT CARD, IDIOT SHEETS

Slang for cue cards.

IMAGE

The projected picture that appears after developing film. Also, any visual interpretation of color, light, or form, whether live, animated, or created for special effect, that is registered on film or tape.

IMAGE DEGRADATION

Loss of clarity and detail in an image (on film or videotape) as a result of too many duplications, technical error, or other modifications.

IMAGE DISTORTION

Change or distortion in the appearance of an object or a scene due to technical difficulty, lens fault, variation in lens focal length, or distortion optics.

IMAGE DUPLICATION

The addition of identical images to different parts of a film frame, using an optical attachment to the camera or projector.

IMAGE INTENSIFIER

An electronic attachment that, when used between the lens and the camera, makes filming possible in very dim lighting conditions.

IMAGE ORTHICON TUBE

An improved iconoscope; a television image pickup in which electrons of low velocity are used for scanning; a television camera picture tube. The Emmy Award was to have been named Immy after this tube until a misprint changed it to Emmy.

IMAGE REPLACEMENT

The replacement of some images on a film with other images by means of special effects processes.

117

IMAGE SPREAD
The slight extension or ghosting of developed silver grains beyond the edges of the photographed images on film formed by light striking the film.

IMAGINARY LINE
An imaginary line between the camera and the subject over which the camera must not cross if one wishes to maintain editing consistency. (See also Axis, Crossing the Line)

IMPROVISE
(1) To ad-lib action or dialogue; (2) to solve an emergency production problem with a unique, creative solution.

"IN"
A directive to begin. "Bring the music in" means to start the music.

IN-AND-OUT-OF-FOCUS EFFECT
See Breathing.

IN-CAMERA MATTE SHOT
A shot accomplished by masking off a part of the field of action by placing a matte in front of the camera lens or at the focal plane. The film is then rewound and reshot with the desired image filling the blocked or matted area. (See also Matte Shot)

IN CHARACTER
(1) Playing a role believably; when a performer portrays the attitude and characteristics of the part he or she is playing without interjecting something from his or her own personality that does not fit the character being played; (2) any dialogue, gesture, costuming, or idiosyncracy that is suitable for a role and would be believably a part of the character being portrayed. (See also Out of Character)

IN CLEAR
Dialogue, music, or sound heard alone. (See also Cold)

IN-HOUSE (Also In-Plant)
Services, talent, or facilities available within a production company or studio. For example, many studios have their own in-house advertising and publicity departments so they do not have to hire an outside or independent source to do the job.

IN-HOUSE UNIT
A production unit or company that is part of the parent company for which it makes feature films or television shows.

IN SYNC
Sound and picture that correspond exactly when played together.

IN-THE-CAMERA EFFECTS (Also In-Camera Effects)
Special effects accomplished by the camera. Changing speed, shooting upside-down, changing the focus, moving the camera while shooting, using mattes, and shooting multiple exposures are examples. (See also Special Effects)

IN THE CAN
(1) Footage that has been shot but not yet processed; (2) a shoot that has wrapped, or completed.

IN TURNAROUND

A property, script, or preproduction plan that is terminated prior to the start of actual production and is therefore in turnaround and available for new negotiations. (See also Turnaround)

IN-BETWEENING

In animation, the creation of intermediate drawings between two key drawings in the sequence of action. For example, if the key action was swinging a bat, the inbetweening would depict the action happening before the swing and the bat actually making contact. (See also Animation)

IN-BETWEENER (Also Assistant Animator)

An animation artist whose job it is to draw the "fill in" action between the principal drawings prepared by the animator. (See also Animation)

INCANDESCENCE

The emission of visible light by a substance at a high temperature.

INCANDESCENT (Also Incandescent Lighting)

A light that uses incandescent (heated tungsten, gas-filled) bulbs or tubes, as opposed to bulbs that use carbon arcs.

INCHES PER SECOND (IPS)

The commonly used term that indicates the measurement of the speed at which tape travels through a tape player.

INCHING

The process of moving film through viewing equipment one frame at a time.

INCHING KNOB (Also Incher)

A knob connected to the drive mechanism of a piece of film equipment that moves the film forward or backward one frame at a time.

INCIDENT LIGHT

Light from all external sources aimed directly at the subject and the light meter as opposed to light that is reflected from the subject toward the light meter.

INCIDENT LIGHT METER

A gauge that measures light falling on a subject by directing the meter toward the light source, as opposed to toward the subject. (See also Exposure Meter)

INCOMING SHOT

The shot immediately following a cut, as opposed to an outgoing or preceding shot; the shot that is next to be seen. (See also Shot)

INDEPENDENTS

Refers to independent filmmakers, producers, or independent stations.

INDEPENDENT FILMMAKER, INDEPENDENT FILMMAKING

Production created, controlled, and conducted totally by individuals not under contract to a major studio. This type of film may use non-union personnel and facilities.

Often an independent producer will try to negotiate a distribution deal with a major source so the film will be seen in the most outlets possible.

INDEPENDENT PRODUCER

A producer who is not under contract to one of the major studios, networks, or production companies.

INDEPENDENT STATIONS

A commercial television station that carries not more than ten hours per week of programming offered by the three major networks in prime time. The remainder of their programming is made up of syndicated or original programming.

INDEPENDENT VIEWFINDER

See Viewfinder.

INDIGENOUS SOUND

Sound from a source visible in the picture, such as a cat meowing.

INDOOR FILM

Color film balanced for incandescent lighting; type A is for light at 3,400 K and type B is for light at 3,200 K.

INDUSTRIAL

An informational film, usually with a plot, made to sell a product or company, or some aspect of the company.

INFINITY

The appearance of unending space in front of the lens; (2) a focusing descriptive or directive to shoot so that the background is clear for as far as the eye can see.

INFORMATIONAL FILM

A film that instructs.

INFRARED

Heat rays; a wide range of invisible radiation; used in certain filming and lighting situations and in the production of infrared mattes.

INFRARED CINEMATOGRAPHY

Filming with special black-and-white or color film that is sensitive to infrared light.

INFRARED FILM

Special film sensitive to infrared. Available in both color and black-and-white, the black-and-white film is usually also sensitive to ultraviolet light, which must be filtered out before shooting.

INFRARED MATTE PROCESS

Making a matte by photographing a performer in front of a screen or background that reflects or transmits infrared light that can be filtered out. The performer's action can then be added to another, already filmed background without the performer's actual background appearing. (See also Matte)

INGENUE

A young actress, usually a teenager or young woman, or a role calling for such an actress.

INKED

Colloquial expression for a signed contract.

INKER
An animation artist whose job it is to draw outlines and details with special acetate ink on the surface of animation cels. (See also Animation)

INKING
Colloquial expression in the animation field that means drawing the outlines for animation artwork. (See also Animation)

INKY-DINK (Also Inkie)
The smallest studio lamp with a bulb up to only 200 watts; it uses incandescent bulbs as opposed to those lamps that use carbon arcs. (See also Dinky Inky)

INPUT
A hole or cable connecting point in equipment, through which electronic signals, for sound or picture, are fed into a monitor, console, recorder, or other device. (See also Output)

INSERT SHOT
A close-up shot of a detail or prop used in the main action to clarify the action, build suspense, heighten the impact, or move the plot along, such as a close-up of a newspaper headline or a gun. Inserts can be shot at any time during production and inserted in the proper place during editing. (See also Cut-away, Reaction Shot, Shot)

INSERT TRAVELING MATTE
A nonmoving matte that has a blank area in a specific part of the frame into which action can be inserted when printed with a complementary matte. (See also Matte)

INSTRUCTIONAL FILM
A film whose primary purpose is to instruct a viewer, i.e., teach skills or attitudes. (See also Educational Film, Training Film)

INT
The abbreviation for Interior, used in a script to indicate that a shot will take place indoors.

INTEGRAL REFLEX VIEWFINDER
A reflex viewfinder found inside a modern film camera that displays key information needed for a shot, including f-stop, VU meter level, and filter indicator, as well as showing the field of action.

INTENSIFICATION
A chemical process used in developing to increase the density and detail of the filmed images.

INTENSITY
(1) The brightness of lighting; (2) the loudness of sound.

INTERCHANGEABLE LENSES
Lenses with standard mounts that can be used on various cameras. (See also Lens)

INTERCOM
A communication system between two or more points with talkback capabilities and a closed circuit sound system, especially used between control booth and studio,

production stage area and offices, and office to office.

INTERCONNECTION
A connecting system through which cable television subscribers in different areas can watch the same programming simultaneously through a microwave or cable linkage of the cable systems' headends. (See also Cable)

INTERCUT, INTERCUTTING
(1) To jump from the picture in one camera to the picture in another, as called by the director; (2) in editing, the insertion of seemingly unrelated shots into a series of related shots for the effect of contrast or other reason. (See also Cross-cutting)

INTERIOR(S)
(1) An indoor set or location; (2) any scene that is shot indoors; (3) shots made outdoors, but presented as having been shot indoors.

INTERIOR CONTINUITY LINKS
The connecting points that bind a nondramatic show.

INTERIOR FRAMING
Use of objects in the foreground to create a center of interest in the shot; framing the shot with props, plants, or whatever appears in the foreground.

INTERIOR LIGHTING
(1) Stage lighting created artificially to give a natural indoor lighting look, such as a living room lit with table and floor lamps. In such cases, the light does not come from the prop lamps, it only appears as if it does; (2) the techniques used in lighting the action for an indoor shoot.

INTERIOR MONOLOGUE
(Also Internal Monologue, Voice-Over)
An audiovisual effect in which a performer is seen and his or her voice is heard, although the performer does not move his or her lips. Used especially to reflect a character's thoughts when reading a letter.

INTERIOR SOUND
The sound of an object as it would be heard from inside it, for example, the sound of a train whistle as heard from inside the train, or the sound of a car engine or horn as heard from inside the car.

INTERLOCK
Any system or arrangement for the playback of picture and sound in perfect synchronization between two units, that is, the projector of the picture and the sound reproducer. The simplest double system has a mechanical link between the audio and video components that allows for synchronous drive.

INTERLOCK PROJECTOR
A film projector that plays the picture while synchronized sound is played on another machine or recorder, which may or may not be a part of the projector.

INTERLOCK SYSTEM
(Also Double System)
A mechanical or electrical system

in which two or more units will start, stop, and play in synchronization. Used with cameras, sound systems, and projectors.

INTERMEDIATE
Any film, other than the camera original, such as a color internegative, or a duplicate positive or negative, that is used to make duplicates.

INTERMEDIATE REVERSAL NEGATIVE
A negative used to eliminate a generation of printing and thus maintain a higher quality by making the negative directly from another negative using the reversal process.

INTERMITTENT (Also Intermittent Movement)
(1) The stop-and-go movement of the film through the gate of a camera, projector, or printer; (2) a discontinuous motion picture movement created by a device such as a Geneva movement or a cam system, used to advance or hold the film in the selected position in the camera, projector, or step-printer apertures; (3) the mechanism itself that pulls the film through the gate a single frame at a time. (See also Gate)

INTERMITTENT MOVEMENT
The mechanism that pulls film in a camera, projector, or printer a single frame at a time. (See also Intermittent)

INTERMITTENT PRESSURE
The pressure on the film in the camera or projector gate when the film is stopped during intermittent movement.

INTERNAL RACKOVER
Used only to set up a shot, this set of sliding mirrors inside a camera intercepts light as it comes through the lens and reflects it into an eyepiece lens. The rackover is not used to monitor action during a shot.

INTERNATIONAL ALLIANCE OF THEATRICAL AND STAGE EMPLOYEES
See IATSE.

INTERNEGATIVE
A duplicate negative, often a composite of A and B rolls, made from a reversal original and used to make release prints to protect the original or source film.

INTERTITLES
Titles that appear within the main body of a film, such as subtitles for dialogue or informational titles, such as dates, and places. (See also Title)

INVERTED TELEPHOTO LENS
See Retrofocal Lens.

INVISIBLE CUT
A cut made during a performer's movement accomplished by using two cameras and switching from one to the other, or by overlapping the action and matching in editing. Invisible cuts make camera shifts less noticeable. (See also Cut)

INWARD THRUST
A motivation or impulse expressed in dialogue or action that moves a character through key plot points.

IPS
Abbreviation for Inches Per Second.

IRIS (Also Iris Diaphragm, Diaphragm)
(1) Located just inside a lens to control the amount of light passing through the lens as calibrated in the terms f-stops or t-stops; (2) a mechanical device used to produce an adjustable circular opening to control light, whether used in front of a camera lens or luminaire; for example, a circular matte created by placing an iris in front of a luminaire; (3) a wipe optical effect in which the wipe line is a circle. The wipe may start in the center of the frame and circle out or at the edge of the frame and circle in. (See also Aperture, Iris Wipe)

IRIS WIPE—IN OR OUT
A circular wipe using an iris. An "iris in" wipe is created by slowly opening a closed iris. An "iris out" wipe is created by slowly closing an iris. (See also Iris, Wipe)

IRRADIATION
Light scattered by silver grains in photographic emulsion. The thicker the emulsion, the more noticeable the effect, which reduces image sharpness.

ISO
Many situation comedies are taped with three or four cameras. The director, sitting in a control booth, selects his "shots" as the play is progressing. In order to guard against mistakes and to provide alternate shots for editing purposes, one or more cameras may be isolated or "iso'd," that is, they are independently hooked up to recording machines. When editing, the director will then have alternative shots to cut into the finished product from the iso cameras.

ITFS
Instructional Television Fixed Service, a microwave frequency band specially allocated for instructional television.

J

JACK (Also Jack Plug)
A male plug connection that joins audio or video components.

JAM (Also Salad)
An occurrence caused by film piling up inside the camera because of mechanical failure or the film itself catching on guide rollers or sprocket wheels. Can also occur with audio tape in a cassette or recorder.

JELLIES
See Gel.

JIGGLE FACTOR
A slang expression referring to the braless bounce (or jiggle) of actresses in certain roles. (See also T&A)

JOY STICK
A stick or handle on a control mechanism to facilitate certain mixes and fades.

JOY-STICK ZOOM CONTROL
A motorized device on a zoom lens, attached by cable to a two-way switch, allowing the lens to zoom in or out instantly. (See also Zoom Lens)

JTC
Script abbreviation meaning "joke to come."

JUICE
A colloquial term meaning "electricity."

JUICER
A colloquial term for an electrician on a set.

JUMP CUT
A sharp, jolting break in a consecutive sequence of action. Three causes of jump cuts are: (1) a mismatch in the action, caused during editing by an overlap or gap between the movement in one shot and the movement in another; (2) too small a change of camera position from shot to shot resulting in the picture looking as if the camera was jerked; (3) too great a change in camera position resulting in the position and direction of the action appearing to reverse itself.

JUNIOR (Also Junior Spot, 2K)
A spotlight with a 1,000 or 2,000-watt bulb; the most commonly used form of studio lighting.

JUSTIFIED CAMERA MOVEMENT
Planned movement by the camera that enhances the action and contributes to the continuity.

JUSTIFIED DOLLY SHOT
See Justified Camera Movement.

K

KALVAR PRINT
A print made by heat process on a Metro-Kalvar printer from a film negative.

KAMMERSPIEL FILM
A term that literally translated means "chamber talk," referring to certain German films made in the 1920s, studio-made and intimate.

KEEP TAKES
Shots that are good and will be used in the finished film.

KELVIN SCALE
The color temperature degree scale; it uses the same degree increments as the centigrade scale but starts at absolute zero. 0° K = -273.16° C.

KEY
See Plate.

KEY GRIP (Also First Grip)
The head or chief stagehand who reports directly to the director of photography. Duties include setting all reflectors, supervising operation of the grip crew in loading, unloading, and placement of all grip equipment such as gobos, scrims, and cookies (see also Cookie), erecting or positioning portable dressing rooms and comfort stations; and pulling cable during moving shots.

KEY LIGHT
The dominant or main light source used to light a subject; usually the brightest, most intense light used in a scene.

KEY NUMBERS
See Edge Numbers.

KEY SOUNDS
The sounds that give an overall impression of the environment being filmed.

KEYSTONING
An image distortion that can occur when the titles are filmed at an improper angle.

KICKER LIGHT (Also Kicker, Cross Backlight, Side Backlight, Hair Light)
A light placed behind and to the side of the subject being photographed. Also, the light created by so placing the lights.

KIDVID
A slang expression that refers to children's television programming.

KILL
A command to turn off the cameras, lights, or sound immediately.

KINESCOPE RECORDING (Also Kinescope, Kine)
A film made by photographing the image from a television screen.

KINESTASIS
A technique of using short shots, only a few frames each, of non-

moving objects such as paintings, or still photographs.

KINETOGRAPH
One of the first motion picture cameras developed for Thomas Alva Edison by W. K. L. Dickson.

KINETOSCOPE
An early motion picture form that could be viewed by only one person at a time peering through a binocular device at a short film loop. Also called "peep shows"; developed for Thomas Alva Edison by W. K. L. Dickson and were usually coin operated.

KINETOSCOPE FILM
A 35mm celluloid-base, flexible loop-film of a single subject, used in a kinetoscope.

KLEIG LIGHT (Also Kleig)
The trade name for a line of lights used in studios, and thus often used to describe any lighting instruments used for lighting purposes in studios. The giant search lights that pan the sky at film premieres are also Kleig lights.

KNIFE SWITCH
A switch with a moveable flat bar or blade, named for its resemblance to a knife.

KUKALORIS
See Cookie.

LAB (LABORATORY)
An establishment that processes and prints exposed film, does soundtrack work, and produces the various masters, inter-negatives, needed workprints, and completed release prints.

LABORATORY EFFECTS
Special effects that can be accomplished in the lab by various printing and processing techniques. (See also Special Effect)

LACING
See Thread.

LACQUER
A specially manufactured liquid used to protect film from scratches. A high-quality lacquer can be removed from the film by use of a solvent and a new coating applied, as often as necessary.

LADIES' COSTUMER
The individual on the crew whose responsibility it is to obtain and maintain the women performers' costumes.

LAID-IN MUSIC TRACK
A music track that has not been edited to the picture but played from its source directly onto the track.

LAMP
Informally, a bulb and its housing; however, in the field of industry lighting, lamp refers to the actual bulb component of a lighting unit, as in a halogen lamp, or a quartz lamp. (See also Halogen Lamp, Quartz Lamp, Light)

LAMP BASE
(Also Light Housing)
A lamp mounting that houses the electrical contact points.

LAMPHOUSE
The part of the projector that houses the lamp or carbon arcs.

LAP DISSOLVE
See Dissolve.

LAPEL MICROPHONE
A small mike that clips onto clothing close to the performer's head. (See also Microphone)

LATE FRINGE TIME
A segment of television viewing time that follows Prime Time; from 11:00 P.M. to 1:00 A.M. Eastern, Mountain, and Pacific Time, and from 10:00 P.M. until midnight in the Central Time Zone. (See also Daytime, Early Fringetime, Family Hour, Fringetime, Prime Access, Primetime)

LATENSIFICATION
Prolonged low-level illumination of a latent image on unprocessed film to increase the shadow area exposure.

LATENT IMAGE
An image that is invisible until the film has been developed and that then is only visible when light is passed through the film.

LATERAL ORIENTATION
The reversed or right-reading position of an image. (See also Right-reading)

LATITUDE
The range to which a film can be under- or over-exposed and still create a satisfactory picture.

LAUGH TRACK
(Also Canned Laughter)
A prerecorded loop or track of laughter to effect or augment live audience laughter, added in the editing or sweetening process.

LAVALIER (Also Neck Microphone)
A small mike worn around the neck on a cord. It may also be clipped under a performer's clothing. (See also Microphone)

LAY (Also Lay-In)
The process of splicing sound tracks to the film in perfect relation to the picture.

LAYING TRACKS
The process of editing sound tracks, ensuring that they are in the correct position for mixing.

LEAD
(1) The principal performing role in a production; (2) time during which the camera is running before the main subject enters the field of action.

LEAD SHEETS
See Bar Sheets.

LEADER
Non-image film spliced onto film preceding actual photography (head leader) or immediately following it (tail leader), used to facilitate threading, editing, or identification. The four most common types of leader are: (1) light-struck—undeveloped film with a gray appearance; (2) black—film developed to a dense black; (3) white—film with a white coating; and (4) clear—film base with no coating. (See also Black Leader, Fill Leader, Film Leader, Light-Struck Leader, Machine Leader, Projection Leader, SMPTE, Universal Leader, Working Leader)

LEAPFROGGING
The practice used by cable systems of skipping over a television station in the cable's geographical area in favor of a signal that is farther away but provides better or more popular programming. The practice is limited by FCC regulations.

LEKOLITE
An ellipsoidal spotlight. A trade name often incorrectly used as a generic term for all ellipsoidal spotlights.

LENS
(1) A precisely shaped piece of glass or plastic that focuses light and projects images onto film, tape, or other surface; (2) a complete opti-

cal system comprised of a lens, barrel, focusing ring, and aperture settings. (See also Anamorphic Lens, Anastigmat Lens, Apochromatic Lens, Close-up Lens, Coated Lens, Diffusion Lens, Diopter Lens, Eyepiece Lens, Fast Lens, Fisheye Lens, Flip Lens, Interchangeable Lens, Long Lens, Macro Lens, Macro-focusing Lens, Macrozoom Lens, Medium Lens, Normal Lens, Plus Lens, Polarized Lens, Retrofocal Lens, Short Focal Length Lens, Split Field Lens, Step Lens, Taking Lens, Telephoto Lens, Viewfinder Objective Lens, Wide-Angle Lens, Zoom Lens)

LENS ANGLE
The lens' angle of view (horizontal or vertical) in degrees.

LENS SPEED
See f-Stop.

LENS STOP
See f-Stop.

LENSER
A colloquial name for a cinematographer or photographer.

LEVEL
Any metered amount of sound or light.

LIBRARY SHOT
A shot from stock footage. (See also Shot, Stock Footage)

LIBRARY SOUND
Music and sound effects from a sound library of prerecorded disks or tapes compiled and catalogued for production use. (See also Sound Effects Library)

LIGHT
(1) Illumination used in filming and taping; (2) the act of arranging illumination for shooting; (3) a unit of lighting containing a lamp and lamp base used for illumination, i.e., spotlight. (See also Spot Light, Lamp, Lamp Base)

LIGHT-BALANCING FILTER
A lightly colored filter, either yellow (warm) or blue (cool) used on the camera lens to compensate for interior lights, weather, or time of day. (See also Filter)

LIGHT BANK
See Bank.

LIGHT BOX
A boxlike device with bulbs inside and a translucent panel on one side used especially as a background for titles or for photographing small objects without light shadows.

LIGHT CHANGES
Printer correction of over- or under-exposed shots, determined by careful examination of each shot, often using an electronic image analyzer. Light changes are often cued with a timing card. (See also Timing Card)

LIGHT END
The final processing area or room into which dry film emerges. This area can be lighted as opposed to the darkened processing end.

LIGHT HOUSING
See Lamp Base.

LIGHT LEVEL
The degree of light intensity as

measured in footcandles (candles per square foot).

LIGHT METER
See Exposure Meter, Incident Light Meter, Spot Brightness Meter.

LIGHT PIPING
Exposing part of the emulsion by piping the light inward from the edge of the film base.

LIGHT PLOT
(1) A diagram of a studio set or location pinpointing the positions and types of lights to be used; (2) a story plot that is not substantive in relation to the action filmed.

LIGHT SOURCE FILTER
A filter fitted over a light in order to change its color or temperature. (See also Filter)

LIGHT STRUCK
Film that has been accidentally exposed to light.

LIGHT-STRUCK LEADER
Intentionally exposed film used as leader. (See also Leader)

LIGHT TIGHT
Any device, container, or area that is constructed in such a way that light can be totally eliminated.

LIGHT TRAP
An entrance into a darkroom designed to allow people but not light to enter. It is usually a two-door hallway, a maze-type walk-way, or a revolving door.

LIGHT VALVE
A mechanism that modulates light for optical sound track recording.

LIGHTING
(1) The illumination and illuminating effects used on the field of action for production; (2) the process of setting the lights for filming or taping.

LIGHTING BALANCE
(1) The range of intensity between the lighting in the foreground and of that in the background; (2) the proper level of lighting to insure that all performers and props in various areas on the set are sufficiently lighted.

LIGHTING BATTEN
A long box or tube with electrical outlets from which lights are suspended.

LIGHTING CONTROL
The control of variation or intensity of light on a set.

LIGHTING CONTROL CONSOLE
A panel of switches by which lights are controlled.

LIGHTING GRID
See Grid.

LIGHTING INSTRUMENTS
(Also Luminaire)
The correct term for the various lighting units, consisting mainly of housing, reflector, and lamp (bulb). Used to light sets for production.

LIGHTING RATIO
The ratio of the intensity of the key light to that of the fill light. (See also Fill Light, Key Light)

LIMBO
A setting with no information about the character or the location. Usually accomplished by using a pure black background or nonidentifiable backdrop and used especially for close-ups and insert shots.

LIMITED ANIMATION
A technique in which only small parts of the whole picture are animated, such as, mouth, eyes, and limited gestures, while the main figures are still. Used mostly to reduce the cost of animation. (See also Animation)

LIMITER
An electronic device that keeps volume, in recording or playback, from exceeding predetermined levels. It can be built-in or added on to audio components.

LIMPET MOUNT
(Also Suction Mount)
A camera mount that allows the camera to be attached to a smooth surface by means of specially designed suction cups.

LINE
A portion of dialogue spoken by a performer that may consist of a word, group of words, or a complete sentence. (See also Lines, One-Liner)

LINE CUTTING
Eliminating dialogue (lines or whole speeches) during editing or any time during production.

LINE PRODUCER
A producer with minimal creative input; one who supervises the overall production coordination under the executive producer.

LINE-UP TONE
A steady one-frequency tone, prerecorded on an audio track for the purpose of matching the volume of other speech, sound effects, or music to be added to the track.

LINES
Dialogue. (See also Line)

LINING UP A SHOT
(Also Lining Up)
The process of choosing the exact camera placement for a shot, usually done by the director and often with the use of a director's finder or the camera's viewfinder.

LINNEBACK
An open-ended box with black interior and small bulb used to view transparencies.

LIP SYNC
(Also Lip Synchronization)
The matched movement of lips and a recorded voice. In musicals another voice is often dubbed in over that of a performer who does not have a singing voice. Exact lip sync must be maintained. (See also Synchronization)

LIQUID GATE (Also Liquid Immersion Gate, Wet Gate Printing)
A printer gate that allows the film to be totally immersed in fluid, preventing and filling in processing

scratches that would appear on the print.

LIVE ACTION
Action by live performers.

LIVE ACTION CINEMATOGRAPHY
The cinematography of live performers, as opposed to the cinematography of animation.

LIVE RECORDING
The process of making an original recording from a live source, as opposed to combining previously recorded tracks.

LIVE SOUND
Sound that is recorded at the moment it occurs and while the action is being photographed, as opposed to previously recorded sound that is being played back. (See also Sound)

LIVESTOCK MAN
(Also Wrangler)
A crew member responsible for the well-being of domestic animals used in a production.

LOADER
See Film Loader.

LOCAL ORIGINATION
Original programming emanating from the source of broadcast.

LOCALE
The location at which a production appears to have been photographed, whether or not filming actually took place in the location represented. For example, matte photography, special effects, and/or studio sets can recreate a designated locale, such as a European town, a desert, or outer space.

LOCATION
A setting or place away from the studio used to film or tape a production. Productions filmed or taped entirely on location used no sets or soundstages at a studio, but photographed their production using the sources and scenes available at the location. A location can be an entire city, a building (interior and/or exterior), a park, a race track, or a forest. (See also Set)

LOCATION BREAKDOWN
(1) A complete breakdown of the scenes and all that is involved (staff, sets, transportation, etc.) in a location shoot; (2) the breakdown of the costs of a location shoot.

LOCATION LIGHTING
Lighting used for filming on location, both exterior and interior, requiring the use of portable electrical equipment.

LOCATION SCOUT
The member of the crew whose responsibility it is to find the locations that suit the needs of the producer and director, as specified in the script. Usually the location scout tours the area of the shoot, taking stills of the various spots considered suitable for the shoot. The director and producer will then discuss with the location scout whether or not the sites are acceptable.

LOCATION SHOOTING
A shot photographed away from a studio set or soundstage, either out-

LOCATION SOUND

doors or in another building. Many productions will film or tape interior shots in a studio and exterior shots on location for a more realistic look.

LOCATION SOUND

Sound recorded live at a location other than a studio or soundstage; location sounds, such as traffic or airplanes are often recorded and stored in sound libraries for controlled use in production. (See also Sound)

LOCKING RING

An adjustable tightening device designed to secure an attachable and/or adjustable piece of equipment.

LOG

(1) A charted step-by-step record of all production schedules and activities; camera and sound logs are usually kept separately (see also Camera/Sound Log); (2) a television station's written record specifying exactly which programs and commercials ran and when aired. The log is usually typed daily by a secretary and kept by the technician or announcer on duty.

LOG BOOK

The book or group of papers used to record all log information. (See also Log, Logging)

LOG SHEET

A page of a log book.

LOGGING

The process of recording itemized shots in a log book by scene and take number, footage and frames used, and edge numbers.

LOGO

A creative concept expressed in graphic design used to represent a company's name, product, or image. A dog peering at a gramophone was a logo used by RCA for many years.

LONG FORM

A type of programming format consisting of productions of one hour or longer including movies for television (as opposed to a theatrically released film) specials, and mini-series.

LONG LENS (Also Telephoto Lens; Long Focal-Length Lens)

A lens with a narrower field of view and longer-than-normal focal length for any frame size; a lens at least 75mm in focal length. A long lens makes subjects at a great distance appear closer to the camera; the closer appearance, relative to the actual distance from the camera, is dependent upon the power or focal length of the lens. (See also Focal Length, Lens, Normal Lens)

LONG PITCH

An extended pitch distance, or distance between film perforations, used especially on prints other than originals, to prevent slippage in the printer or projector. Also used with films in high speed cameras. (See also Pitch)

LONG SHOT (L.S., Also Establishing Shot)

A shot that includes the full figure of a subject, the foreground and background, thereby allowing a relatively full view of the set or location; a shot made a considerable distance from the subject. (See also Shot)

LOOP (Also Loop Film)
(1) A strip of film or tape joined to itself to form a circle, used for any repetitive purpose, e.g., duplication, multiple projections, looping, or for recorded sustained effects such as traffic noise or crowds; (2) the slack of film on either side of a film gate or sprocket that expands and diminishes dependent on the pull of the camera mechanism or projector movement, the loop allows for smooth transition from intermittent to continous movement (see Film Loop); (3) the act of looping. (See also Sound Loop)

LOOPING
The process of rerecording dialogue to match the lip movement already filmed with a noisy or faulty sound track. So named as the lip-synching is done to a loop film that repeats itself until the performer matches the screen lip movement exactly; most dialogue recorded on location is looped because such background noises as the sounds of a busy street cannot be controlled.

LOOPING STAGE
See Dubbing Stage.

LOOP TREE
The framework upon which film or audio loops are revolved, including sprockets and rollers that repeat the film or tape as desired.

LOW BUDGET PRODUCTION

LOOSE CLOSE-UP
See Medium Close-Up.

LOOSE GATE
A projector or printer gate that has been deliberately set looser to prevent the possibility of film damage on certain films.

LOOSE SHOT
A shot framed so that there is considerable space between the subject and the edge of the frame. (See also Shot)

LOT
A film or television studio and the grounds associated with it where production takes place. The lot houses sets, scenery, and props, as well as soundstages for filming and taping. (See also Studio Lot)

LOUDSPEAKER
See Bullhorn.

LOW ANGLE
A camera positioned below and pointed up toward the subject.

LOW-ANGLE SHOT
A shot made from a low angle, i.e., shooting up at the subject. This angle has the effect of enlarging the size of the subject. *Land of the Giants* is an example of a show shot predominantly from a low angle. (See also Camera Angle, Shot)

LOW BUDGET PRODUCTION
A production that must operate on restricted funds, therefore using less expensive production techniques.

LOW CONTRAST
A long gradation of black and white tones in a photographic image; not a sharp distinction between the blackest black and whitest white, a softer, more gray-toned image. (See also High Contrast)

LOW CONTRAST ORIGINAL
An original reversal film from which prints with good projection contrast are made.

LOW KEY LIGHTING
(1) Lighting that is subdued and mellow to create a soft look; (2) a lighting style emphasized by dimly lit, dark-toned backgrounds, dark costumes and sets; (3) a minimal level of illumination of the subject. (See also High Key Lighting, Light)

LOW-NOISE LAMP
A specially designed lamp possessing little or no electrical buzz or noise, used mostly in audio systems.

L.S.
Abbreviation used for "long shot."

LUBRICANT
Any protective substance applied to film to reduce friction when the film is passing through projection or photographic machinery.

LUMEN
A measurement of light equal to the light of one standard candle cast on a surface of one square meter and at a distance of one meter from the candle. (See also Lux)

LUMINAIRE
The proper, general term for a complete light instrument (housing, lamp, and cable); also covers all types of lighting units used in production. (See also Lighting, Instruments)

LUMINANCE
The measure of actual brightness emanating from a light source, as opposed to the sensation of brightness that can be affected by surrounding atmosphere or color.

LUMINESCENCE
See Video Up.

LUX
One lumen per square millimeter.

M

MACGUFFIN (Also Maguffin)
Coined by Alfred Hitchcock, the MacGuffin is something that motivates the characters and their actions but need not be of interest in itself to the audience. It is a plot element or gimmick that is only necessary to move the action. In a spy thriller, the MacGuffin could be secret documents that the villain is intent on smuggling out of the country. The documents need not be described in detail and their purpose may not be clear to the audience, but they motivate the characters nonetheless. In Hitchcock's *The Thirty-Nine Steps*, the MacGuffin was the formula for an airplane's construction.

MACHINE LEADER
A strip of strong leader film used to guide and pull film through a film processing machine. (See also Leader)

MACRO-FOCUSING TELEPHOTO LENS
(Also Macro-Telephoto Lens, Macro-Tele Lens)
A telephoto lens capable of taking extreme close-up shots. (See also Lens)

MACRO LENS
A lens with the capability of focusing on subjects positioned close to the camera. (See also Lens)

MACROCINEMATOGRAPHY
A technique of filming of extremely small objects, such as insects.

MACROZOOM LENS
A zoom lens with the capability of focusing on objects very close to the camera. (See also Lens, Zoom Lens)

MAGAZINE
A light-tight container that houses and allows for the dispensing and take-up of film, used in some cameras, printers, and optical sound recorders.

MAGAZINE BARNEY
A padded cover that when used to cover a camera magazine muffles its sound.

MAGIC HOUR
Dusk or dawn, a time when the sun casts a particular light that gives a surrealistic quality to a shot.

MAGENTA
(1) One of the primary color reproduction shades used in film; a reddish purple. The other primary film colors are cyan and yellow; (2) a commonly used stage lighting color, the color that occurs when red and blue lights are mixed, or when a magenta gel filter is used.

MAGNETIC FILM (Also Mag Film, Full Coat)

Standard width film, with sprockets on one side for 16mm and on both sides for 35mm and 70mm, with a coating of iron oxide compound on which sound can be recorded and played back.

MAGNETIC FILM RECORDER

A sound recorder using perforated magnetic film, as distinguished from a magnetic tape recorder.

MAGNETIC HEAD

An essential component of any audio, video or film unit that uses magnetic film or tape. It operates in contact with the iron oxide coating on the tape or film, records, plays back, and erases sound or video.

MAGNETIC MASTER

(1) The final edited and mixed magnetic sound recording from which the release print soundtrack is made; (2) a magnetic track that will be added to or mixed into the final edited master. (See also Mix)

MAGNETIC ORIGINAL

An original soundtrack recorded live on magnetic film or tape.

MAGNETIC OXIDE

An iron oxide, the magnetically sensitive substance used to coat recording tape and sound film.

MAGNETIC RECORDING

A technique of sound recording whereby magnetic patterns representing the sequential wave patterns of the sound being recorded are imprinted on the magnetic strip, film or wire used for recording.

MAGNETIC SOUND TRACK

Audio recording in which the sound is interpreted as variations in a magnetic field that runs, usually in a strip, the length of magnetic film or tape. (See also Magnetic Recording, Magnetic Stripe, Magnetic Tape)

MAGNETIC STOCK (Also Mag Stock)

Commonly used term that refers to magnetic film as packaged by a manufacturer. (See also Magnetic Film)

MAGNETIC STRIPE (Also Magnetic Stripe Sound, Mag Stripe)

A strip of iron oxide capable of recording and reproducing sound, glued along the edge of clear film or release prints in the standard sound track position, used especially for television news. A mag stripe, as it is most commonly called, is used to record single-system newsreel sound tracks and to produce the soundtracks on super 8mm release prints. (See also Sound Stripe)

MAGNETIC TAPE (Also Audio Tape, Mag Tape, Quarter-inch Tape, Sound Tape)

A thin plastic tape coated with magnetically sensitive iron oxide on which magnetic patterns become sequential wave patterns of sound, thus having the capability to record and reproduce sound.

MAGNETIC TAPE RECORDER

An audio recording machine that records and plays back sound on magnetic tape.

MAGNETIC TRACK

An audio tape with magnetic patterns that can become sound when played back on a magnetic tape recorder.

MAGNETIC TRANSFER

Any transfer, usually for editing purposes, from one magnetic medium to another, as from magnetic tape to perforated magnetic film.

MAGNETIC WORKPRINT

Magnetic soundtrack that has been rerecorded, usually on perforated film with coded edge numbers; used in editing and recording.

MAGOPTICAL TRACK
(Also Magoptical)

A soundtrack with both optical track and magnetic stripe track. By combining four magnetic tracks with an optical track, this type of film track can be used in virtually every kind of theater from those with simple monaural optical capabilities to those with four-track playback capabilities.

MAGOPTICAL RELEASE PRINT

A release print comprised of both optical and magnetic soundtracks.

MAGUFFIN

See MacGuffin.

MAIN TITLE

The combined credits and titles that run at the opening and at the close of a filmed or taped production. The word "credits" seems to be used more in association with a taped television production, and the word "titles" are more often associated with films. (See also Closing Credits, Credits, Opening Credits)

MAJOR MARKET (Also Top-100 Market)

A term that refers to one of the one hundred largest television broadcast markets, ranked by the number of viewers. Markets are rated from 1 (New York) to 100 (Lansing, Michigan).

MAKEUP, MAKE UP

(1) Makeup: the cosmetics applied to a performer's face, and sometimes body, to enhance the appearance or to create a character; (2) make up: the act of applying makeup. Modern film makeup can be so extensive for certain character roles, especially aging a young performer or for science fiction and horror roles, that one performer's makeup can take five hours or longer to apply; (3) the studio or production location where makeup is kept and applied; the makeup department; (4) "make-up"—the command to bring makeup to the set, usually for a touch-up, especially powder; (5) a rescheduled spot announcement, scene, or performance. (See also Character Makeup, Cosmetic Makeup, Grease Paint Pancake, Straight Makeup)

MAKEUP ARTIST
The crew member whose responsibility it is to design and apply the cosmetics and character makeup worn by the cast members.

MAKE-UP CALL
The time when a performer must report to the makeup department before appearing on the set.

MALTESE CROSS MOVEMENT
See Geneva Movement.

MANUAL DIMMER
A hand-operated device used to dim or control the intensity of lights.

MANUAL OVERRIDE
(Also Override)
A mechanism bypassing the automatic control that holds open the iris of a camera. This control is found on cameras that have built-in light meters that automatically control the iris of the lens.

MARKER
See Slate.

MARKS
See Blocking.

MARQUEE
The large, usually neon-lit sign in front of a theater upon which the title of the production and often the stars' names or other saleable information is displayed in large, changeable letters.

MARRIED (Also Composite Print, Married Print, Wedded Print)
A film strip in which the picture and soundtrack have been combined.

MASK
The means by which certain parts of a film are prevented from being exposed or from appearing on the final print, while the remainder of the film is undisturbed. The mask can be: (1) a treated strip of film traveling in contact with the film to be exposed; (2) a custom-sized and-shaped plate that is added to a camera, projector, or printer; (3) a custom-sized and shaped device placed in front of a lens. A mask is used often to obtain a special effect for filming a scene. At an outdoor location the modern background can be masked out and in processing a matte painting of a 1920s background can be added. *The Sting* used this process. (See also Matte)

MASKING
(1) The process of using a mask to block certain parts of the photography and prevent their appearance on the final print; (2) the black, nonreflective border that surrounds the perimeter of a motion picture screen. Masking in motion picture theaters can be adjusted to cover more or less of the screen, depending upon the lens or size of the film. Most theaters today electrically control the masking.

MASTER
(1) The original film, videotape, or

audio tape recording on which all elements of editing and special effects have been finalized; the final film or tape; (2) a strip of film, or piece of videotape or audio tape, that can be one or more generations away from the original footage on which effects or additional elements have been added or combined for inclusion in the final film or tape; (3) a tape track on which other tracks have been combined.

MASTER LONG SHOT
See Master Shot.

MASTER POSITIVE (Also Fine Grain Master Positive)
A timed print made from a negative original in which optical effects and/or any footage from optical printing can be included. "Master Positive" usually applies to 35mm production. (See also Timed Workprint)

MASTER SCENE SCRIPT
A script with scenes listed and numbered but without shot breakdowns.

MASTER SHOT
A long shot or moving shot encompassing all of the action in a particular scene and giving an overall view of the action. Sometimes called an establishing shot, a master shot establishes all the important scene elements in the viewer's mind. (See also Cover Shot, Establishing Shot, Reestablishing Shot, Shot)

MASTER SHOT TECHNIQUE (Also Master Scene Technique)
A filming technique used by the director to permit easy match-action editing and continuity without jump cuts. The action is first filmed with a long shot, then repeated exactly and filmed with medium shots, and again with close-ups. If several cameras are available the action can be shot simultaneously from the various angles used. (See also Single Shot Technique, Shot)

MASTER SWITCH
See Blackout Switch.

MASTERING LAB
A production facility with the capabilities of producing a master recording from which duplicates can be made.

MATCH ACTION CUTTING
See Matching.

MATCH DISSOLVE
A dissolve connecting images of similar content or form. Examples are dissolving the image of an eagle taking off into an airplane taking off, or bouncing balls becoming bubbles, or the image of a child running into the image of a running athlete.

MATCH-IMAGE CUT
A cut from one image to another image with the same basic shape as the one in the preceding shot, such as cutting from a toy truck to a real one, or from a bird to an airplane.

MATCHING

In editing, the process of conforming (cutting) the original film to match the workprint. This technique allows final prints and dupes to be made from the camera original, thereby ensuring dupes of the highest quality. (See also Workprint)

MATCHING ACTION

In editing, the process of choosing and combining two pieces of action at a point where the action will smoothly overlap so as to achieve a smooth cut. (See also Overlapping)

MATRIX

One of the three dyed-emulsion image strips on the film base that produces color on film.

MATTE

(1) A specially designed mask with one or more specified areas cut out of it that, when placed on the camera or printer, allows for the exposure or filming of only the areas not blocked. Live actors can be filmed on a modern street in 1920's clothing, with a portion of the modern building matted out. In editing, a matte painting of a 1920's background can be inserted to fit into the matted area, thereby giving the appearance of a 1920's setting. Mattes are most commonly mounted in front of the camera lens during photography or immediately in front of the film in a printer. Mattes can be cut to exact specifications to block as much or as little light as desired to restrict the image to a corresponding part of the film frame; (2) the process of using mattes. When mattes are set up as film rolls synchronized with a film being duplicated to block unwanted images from being duplicated, they are traveling mattes or matte rolls; (3) any device attached to the front of a light source or camera, projector, or printer lens to shape the light beam. (See also Composit Matte, Counter Matte, Dunning-Pomeroy Self-Matting, Garbage Matte, Infrared Matte, Insert Traveling Matte, Mask, Photo Matte, Special Effects Matte, Traced Matte, Traveling Matte, Ultraviolet Matte)

MATTE BLEED

A technical difficulty with a matte or matted image that makes the matte lines noticeable on the film.

MATTE BOARD

A frame support or easel that supports mattes while they are being photographed.

MATTE BOX

An attachment for the front of a camera lens that has slots to hold mattes, filters, and a sunshade.

MATTE GLASS

See Flat Glass.

MATTE-LINE

See Blend-line.

MATTE PAINTING

A painting used as background, to fit exactly as a matte has been cut. The action is filmed against a background that has been matted out; the painting is inserted, possibly by traveling matte in editing, as the correct background. In the film

Earthquake matte paintings were used as background depicting the destruction of Los Angeles. Matte paintings can also create the illusion of having been filmed in a foreign country. For example, action is filmed on a soundstage and in editing matched with a matte painting of the Eiffel Tower and the skyline of Paris, or whatever locale is called for in the script.

MATTE SCREEN
A flat, nonbeaded screen, used in screening rooms where viewers will be seated at the sides of the screen. A matte screen reflects a nondistorted image to those viewers seated at the side.

MATTE SHOT
Any shot in which a portion of the field of action has been blocked with a matte for other action or background to be added later, either in the original camera, in a bipack camera, or in an optical printer with a traveling matte. (See also In-Camera Matte Shot, Matte, Matte Painting, Shot, Traveling Matte)

MAXI BRUTE
See Brute.

MAXIMUM APERTURE
The widest opening of a lens on a camera or printer head that exposes the greatest amount of light.

MCU
Abbreviation for Medium Close-Up.

MEAL PENALTY
A fine levied by the unions on a production company. The company must pay crew and performers an additional sum because a predetermined number of hours elapsed without a meal break.

MEAT AXE
A casual term for a small flag or scrim holder.

MECHANICAL SPECIAL EFFECTS
Fires, explosions, pyrotechnics, avalanches, storms, or similar effects created by artificial means, usually during principal photography. (See also Special Effects)

MECHANICALS
The actual disc recordings as made from a mixed audio track.

MEDIA
The available sources of mass communications, such as television, radio, and newspapers, treated as a group and referred to as the media, the plural of medium. It has come to include any or all of the mass communication sources.

MEDIUM CLOSE-UP
(Also Loose Close-Up)
A shot of only a person's head, shoulders, and part of the chest. A medium close-up does not allow room for hand gestures as would a medium shot. (See also Talking Heads)

MEDIUM LENS
The mid-range lens that defines the average or normal focal length for the format being used.

MEDIUM LONG SHOT
(ML, Also Full Shot)

A shot in which the camera frames the entire subject, but photographed from a point somewhat nearer the subject than if a long shot were taken, that is, the subject's figure fills or nearly fills the frame. (See also Shot)

MEDIUM SHOT
(MS, Also Half Shot)

A person or object is framed roughly between a close-up and a long shot; if a person (or persons), the shot is framed from the waist up allowing visible hand gestures; if an object, the shot is framed from the approximate center up. (See also Shot)

MEGGER

The person who held the megaphone as the director called out instructions to the performers in the early days of filmmaking.

MELODRAMATIC FILM

A film whose plot focuses on suspense and action, using broad physical action; usually with a happy ending. Characterizations in such a film are not usually developed to any great depth.

MEMORANDUM AGREEMENT

See Deal Memo.

MEN'S COSTUMER

The person hired by the production company to secure and maintain the men's costumes.

MERCER CLIP

The brand name of a commonly used small plastic clip that holds the ends of a film intact during editing.

METER READING

The selection of the correct f-stop on the lens, determined by measuring the intensity of the key light with a light meter; the act of measuring the intensity is referred to as meter reading. (See also Exposure Meter)

METHOD ACTING (Also Method; Stanislavski Method)

An acting style developed and made famous by Stanislavski, of the Moscow Art Theatre, whose name is synonymous with the technique in which the performer is instructed to become the character, not play the role. The actor derives characterization by drawing on personal experience that can parallel what the character is feeling.

METTEUR-EN-SCENE

The person responsible for the mise-en-scene or the final appearance of the film including lighting, sets, costumes, performances, and editing. Usually the director. (See also Mise-en-scène)

MICKEY MOUSE

Predictable and obvious sound, music, and/or effects in exact correlation with the action, as in cartoons. Examples are a slow, loud drum beat to correlate with slow, heavy footsteps, or music playing down a scale as a cartoon figure slides down a bannister.

MICROPHONE (Also Mike, Mic)

(1) An electrical instrument that carries sound from its source to a

recorder, amplifier, or sound reproduction unit (see also Cardioid Microphone, Crystal Microphone, Directional Microphone, Label Microphone, Lavalier, Nondirectional Microphone, Off-Microphone, Omnidirectional Microphone, Radio Microphone, Radio Wireless Microphone, Unidirectional Microphone); (2) the command, for example, "mike the piano," means to increase the sound level, usually by adding a microphone.

MICROPHONE BOOM

A long pole to which a microphone is attached. The boom is then suspended over the field of action, outside of camera range.

MICROPHONE INPUT

A hole or connecting point on an audio unit in which a cable is inserted for the purpose of connecting a microphone to the unit. (See also Input)

MICROPHONE PICK-UP PATTERN

The spatial area in which a microphone has its most effective sound-receiving sensitivity. (See also Unidirectional Microphone)

MICROPHONE PLACEMENT

Selecting the placement of one or more microphones that will result in the most effective sound pick-up.

MICROPHONE PRESENCE

The nearness and clarity of a speaker's voice when talking into a microphone. When too close to the mike, breathing and undesirable lip sounds become audible and the speaker has too much presence.

MICROPHONE SHADOW

The unwanted shadow created by the microphone on the set.

MICROWAVE RELAY

The transmission to cable systems of television signals from distant locations in which the signals are received directly over-the-air.

MID-SEASON REPLACEMENT

A television series not broadcast at the beginning of a network's new TV season but held in reserve to air in place of a series that is cancelled in mid-season, or at any time after the new season has begun.

MIDT OUT SOUND

See MOS

MIKE MAN

The crew member whose responsibility it is to place the microphone for filming.

MIMEO

Studio department where final scripts are duplicated for distribution to the cast, production executives, and all involved in that script's production.

MINI BRUTE

See Brute.

MINIATURE (Also Model)

A scale model of an object or set. Miniatures are used in a film primarily to reproduce catastrophes, such as train wrecks or explosions. They were used in the original film *King Kong* and the remake. Today, they are used in disaster, fantasy, and science fiction films such as

Star Wars to create realistic scenes, such as spaceship battles, that would be extremely costly and time-consuming to construct if full-size models were required. Today computer-created effects are also used in place of an actual model. (See also Special Effects)

MINIMUMS (Also Scale)

The minimum daily or weekly amount a contractee, such as performer, writer, or crew member, can receive from a production for services rendered. Established by industry-wide arrangement among the guilds, unions, producers, and production companies.

MINOR CHARACTER

A small role or character in a film.

MIRROR BALL

A ball with many small mirrors attached to its surface. The ball is suspended and revolved as spotlights are beamed at it, producing hundreds of moving flashes of light around a room or set.

MIRROR IMAGE

The image reflected in a mirror as photographed by the camera.

MIRROR MOUNT

A firm support that holds a mirror or pellicle used for front projection of a background.

MIRROR REFLEX SHUTTER
(Also Mirror Shutter)

A silvered shutter used to eliminate viewfinder parallax. When the mirror shutter is closed, light coming through the lens is reflected into the viewfinder. Reflex shutters are at 45° to the lens axis.

MIRROR SHOT

(1) Photographing the reflection of an object or action in a mirror or by using a semitransparent mirror to achieve a ghost effect. (See also Shot)

MISCAST

A term that indicates the right performer was not selected for a role; a mistakenly cast actor who has been chosen for a part but through either lack of proper qualifications, wrong "chemistry" with the other actors, or other technical or personal reasons does not perform to par.

MISE-EN-SCENE

The total effect of the elements in the action field, such as costumes, settings, and lighting. (See also Metteur-en-scène, Staging)

MISSILE TRACKING CAMERA

A camera used to photograph the flight of a missile, mounted on a special electrically operated mount to facilitate tracking.

MITCHELL

Previously the industry standard camera and named after its inventor. It has been largely replaced by the Arriflex camera.

MIX (Also Mixing)

Blending the sounds from several tapes, microphones, or tracks onto a single track, including volume adjustments, fades, cross fades, and equalizations of each track. Mixing combines all the elements of a soundtrack recording and is the most important step in completing the final product. In studios that have 24 tracks (48 elemental positions if working in stereo), it often takes hours and sometimes days to have a sound mixed properly for the desired quality and tonal effects. Music can have trumpets playing in the background and at a certain dramatic point the trumpet is dominant; this increase in the trumpet's volume is controlled in the mix. (See also Background Sound)

MIXER (Also Dubber)

(1) An electronic device or console that allows for the combining of several sound sources, such as voice from one track with music from another track. In modern 24-track recording studios, the consoles are capable of mixing the sounds from 24 different sources. (See also Mixing Console); (2) the technician who controls the console for mixing. (See also Sound Mixer)

MIXER BOARD

A sound-mixing console.

MIXING CONSOLE
(Also Console, Mixer)

A desklike board with controls that when set as desired create the desired effects on tape. The console is connected to microphones, tape recorders, and special units, such as an echo chamber. The controls include switches or dials for volume, equalization of treble and bass, special filters and sound devices for each track, which on a 24-track console means 24 sets of multiple controls. The console itself also has patch panels for connecting cables from tape recorders, microphones, instruments, and special units, as well as VU meters to indicate volume levels and switches for remote control of any connected recorders or microphones.

MIXING CUE SHEET

A sheet made up of several columns, one for each track, in which notes will be made regarding elements on each track such as footage, fades, volume levels, and equalizations. Used for mixing sound-tracks. (See also Cue Sheet)

MIXING ROOMS

Production facilities with boards and consoles for the purpose of mixing. (See also Mixing)

MLS

Abbreviation for Medium Long Shot.,

MM

Abbreviation for millimeter. Used for film size. The most common sizes are 8mm, 16mm, 35mm, and 70mm.

MOBILE MICROPHONE

See Wireless Microphone

MOBILE UNIT

(1) A portable camera and sound crew that can film on location, often with a backup mobile van with mini-production capabilities, on news shows or on location production shoots; (2) a van or motor home

equipped with a control booth for on-location shooting or recording, usually used in videotape production.

MODEL
See Miniature.

MODELING
A lighting technique that gives depth and texture to a two-dimensional subject by precise adjustment of the key and fill lights.

MODELING LIGHTS
The key and fill lights that create the highlights and shadows on the actors or furnishings on the set, rather than set lights.

MODULATION
The varying degrees of sound frequencies or amplitudes; a change in pitch or volume of sound or soundtracks. Also, the manifestations of such changes as they appear on an optical soundtrack.

MOLEFAYS
Lights from one-bulb to twelve-bulb clusters manufactured by the Mole-Richardson Company, usually coupled with a FAY type 650-watt bulb, hence the name. This type of lighting is primarily used to provide even illumination in outdoor filming. (See also FAY Lights)

MOLEPARS
One- to nine-bulb light clusters manufactured by Mole-Richardson Company that use 1,000-watt PAR lamps, used in outdoor lighting situations where more intensity is needed than provided by Molefays. (See also Birdseye)

MONITOR
A screen set up independent of the action to view action being taped or filmed. In television studios, monitors are usually found in the control room or booth to aid the director, producer, assistant director, and/or technical director during the taping of a show; in dressing rooms and green rooms for use by cast members who are waiting for their cues; in front of the audience, suspended from the ceiling, to enable the audience to see the action taking place on the set when the cameras are blocking their view; (2) to exercise continuous control over picture and/or sound as they are being recorded.

MONITOR SPEAKER
A loudspeaker used primarily in rehearsals, recording, and mixing so that sound can be monitored during performance and adjusted in the master speakers or sound consoles if necessary.

MONITOR VIEWFINDER
(Also Monitoring Viewfinder)
A small viewing screen on which a film can be viewed without an eyepiece lens, used when the film cannot readily be projected onto a larger screening device. (See also Viewfinder)

MONOCHROMATIC
The range of tones from white through grays to black in black-and-white photographic images.

148

MONOCHROME
Photography in black and white.

MONOPACK
Any color film system that uses only a single strip of film.

MONOPACK SUBTRACTIVE COLOR PROCESS (Also Single Film Subtractive Color Process)
A process in which color film creates colors on the screen. White light is filtered through the film's emulsion layers, which are dyed in various intensities of the primary color reproduction tones of cyan, yellow, and magenta.

MONOPOD
A camera head support with one leg.

MONTAGE
(1) A series of shots or scenes that appear unrelated but when edited together present a unified story; (2) the assembly or editing of a series of shots, often with superimposition and optical effects, showing a condensed episode of events, such as a montage of war scenes from World War I war planes to modern spacecraft, or a montage of the effects of a heat wave on a city. The latter are often used by news shows; (3) a short, impressionistic sequence that shows the passing of time or establishes the characteristics of a scene, location, or period; (4) a colloquial term loosely used to mean editing.

MOOD LIGHTING
Lighting designed to create a mood; psychological lighting such as using dim lighting in a sad scene.

MOOD MUSIC
Background music designed to direct the emotions of the audience and elicit the response intended by the script and the director.

MOS (Also Midt Out Sound)
Without sound; filming silently without sound or track. The phrase was coined in the early days of film when a famous German director instructed the crew to film "without sound," which was pronounced in his heavy accent "midt out sound." European technicians also used this pronunciation and when those crews mingled with American crews, the phrase stuck and is still used today.

MOTIF
A musical theme or passage that is repeated throughout a film.

MOTION PICTURE ASSOCIATION OF AMERICA
See MPAA.

MOTIVATION LIGHTING
Lighting effects whereby the audience accepts that the light comes from a believable source, such as a table lamp, instead of from the actual stage lighting used to create the effect.

MOTION PICTURE
Images that are photographed to appear to move when successions of still frames are pulled quickly by mechanism through a projector; col-

loquially, a film, most commonly a feature film.

MOTION PICTURE ASSOCIATION OF AMERICA
See MPAA.

MOTION PICTURE CAMERA
See Camera.

MOTION PICTURE FILM
See Film.

MOTION PICTURE PROJECTOR
See Projector.

MOVIE
See Motion Picture.

MOVIE FOR TELEVISION
(Also MOW)

A long form of television in which a feature length presentation is aired on one of the major networks. The term was coined from a network format that showed a feature film, either a viewing of a theatrical release or an original film made for television, one night a week. The term has come to mean all original feature-length films made for the television viewing audience.

MOVIE PALACE
A theater that is large and intricately ornate.

MOVING BACKGROUNDS
Action appearing on screens projected behind the field of action and, when photographed, appearing to be in the field of action. Actors can be seated in a stationary car on a set, the car is bounced in place, and a moving background makes it appear as if the car is actually moving.

MOVING CAMERA
A camera that moves in any direction or for any purpose during a shot. (See also Moving Shot)

MOVING CAMERA MATTES
Mattes specifically designed for each frame of those shots filmed with some camera movement.

MOVING SHOT
(Also Traveling)

A shot accomplished while the camera is in motion, either on a dolly, crane, or camera car. (See also Moving Camera, Shot)

MOVIOLA
An upright film-viewing machine in which separate rolls of sound and picture film are played in sync. Used in the editing process to cut, view, and sync sound pictures. Moviola is actually the brand name, but because the machine was the original industry standard for editing, the name stuck. Although Moviolas are still in use, flat bed editing machines are now used extensively in editing.

MOW
Abbreviation for Movie of the Week. (See also Movie for Television)

MPAA
The abbreviation for Motion Picture Association of America. The major film distribution companies have formed this organization

through which problems and opportunities of the industry as a whole can be discussed. The association represents not only distributors and exhibitors but also the interests of the public and business communities. The MPAA is best known by the general public for the rating code given each film with major distribution. These codes are indicated by the letters G, PG, PG-13, R, and X. (See also MPAA Code)

MPAA CODE
(Also MPAA Rating)

An indication of the content suitability for specific audiences of films with major distribution. The code is a letter or letters that appear on all advertising and promotion for the film, and are as follows: G—General Audiences, suitable for an audience of all age groups; PG—Parental Guidance Suggested, some material, usually language, sex, or violence, may be a bit strong for a very young or conservative audience; PG-13—Parents are strongly cautioned to give special guidance for attendance of children under 13; some material may be inappropriate for young children; R-Restricted, children under seventeen are not permitted to view the film without being accompanied by a parent or guardian. These films have stronger language, violence, or sexual content than films with a PG rating; X—Indicates that a film is for adults only and that no one under seventeen is admitted under any circumstances. In most cases this rating is given because of extremely explicit sexual content. (See also Classification)

MS
Abbreviation for Medium Shot.

MULTIPLANE
The photographing of animation artwork at more than one plane in the camera field.

MULTIPLE CAMERA TECHNIQUE
Simultaneous photography using more than one camera. It gives the director the choice of camera angles with a selection of lenses and shots when the film is intercut in editing.

MULTIPLE EXPOSURE
A technique used for special effects. The film, after initial exposure, is rewound and reexposed in the camera. Multiple images can also be added during printing.

MULTIPLE-FRAME PRINTING
A printing technique whereby one frame of motion picture film is printed several times in succession.

MULTIPLE-HEAD PRINTER
A printer that has individual heads for the A roll, B roll, and soundtrack.

MULTIPLE IMAGE
Having several nonsuperimposed images within one motion picture frame.

MULTIPLE PRINTING
A printing technique in which images taken from several different film strips are combined onto one piece of duplicating or print stock.

MULTIPLE SCREEN PRESENTATION
A screening technique in which two or more films are projected on two or more screens, or on sections of one large screen.

MULTIPLE SYSTEM OPERATOR
A company owning more than one cable television system.

MUS
Abbreviation for Music.

MUSIC CUE SHEET
A list of musical cues in a production; a breakdown of the songs or pieces of music played during each specific scene or production number. Used by sound and lighting crews, the producer and director, and by any live musicians who are adding their part to a soundtrack or performing in a live production. (See also Cue Sheet)

MUSIC ROUTINE SHEET
A listing of the musical numbers and acts, along with the performers, in order of appearance.

MUSIC VIDEO
A new genre of audiovisual programming in which a song is accompanied by a film that interprets it and by increasing its entertainment value helps to promote the song and the artist. Although started as experimental broadcasting and mainly performed by rock and roll groups, music videos have become extremely popular and are now played as special and fill programming on networks and major cable outlets; in addition, music videos have now been recorded by such non-rock entertainment superstars as Frank Sinatra and Dean Martin.

MUSICAL (Also Musical Film, Musical Comedy)
A feature film whose plot is enhanced by singing and, usually, dancing, performed by the characters in the film. Often the plot itself is secondary to the musical production numbers. Musicals are not necessarily comedies; one recent example being Barbra Streisand's *Yentl*. Some classic examples of musicals include *Singing in the Rain*, *My Fair Lady*, and *The Sound of Music*.

N

NAR
Abbreviation for the word narration.

NARRATION (Also Commentary, NAR)
Material read by a performer, usually off-camera, to clarify action, move the plot along, add expert opinion, or explain what is being shown. (See also Voice Over)

NARRATION SCRIPT
The script written for a narrator and read from in a voice-over recording session.

NARRATIVE FILM
Generally, a dramatic, not comedic, story film.

NARRATOR
One who relates information about the story that is unfolding on the screen.

NARROW-GAUGE FILM
Any film size smaller than 35mm.

NATIONAL BROADCASTING SYSTEM
NBC, one of the three major U.S. television networks, which has over two hundred affiliated stations.

NATIONALS
The weekly ratings of the top sixty-four television shows appearing on the three networks, listed according to their standings for the week, i.e., number one is the highest rated show. (See also Ratings)

NATURAL COLOR LAMP
A colored bulb that is transparent.

NATURAL LIGHT
Usually refers to sunlight, as opposed to artificial light from lamps.

NATURAL LIGHTING
Light that is present on location from its source without addition of studio lights, such as sunlight, candlelight, car headlights, or street lights.

NC
An early model of the Mitchell camera. Because it is noisy, it is used either with a barney or for shooting wild shots without sound. (See also Wild Picture)

NECK MICROPHONE
See Lavalier.

NEGATIVE
Film that is exposed in the camera and that when developed has reversed images, i.e., light tones are dark. (See also Negative Image)

153

NEGATIVE COST
Various costs associated with the acquisition, development, and production of a film. (See also Above-the-Line Costs)

NEGATIVE CUTTING
See Matching.

NEGATIVE FILM
Film that produces images whose tones are reversed. (See also Negative Image)

NEGATIVE IMAGE
A photo image whose tone is reversed from that of the live scene. What was light in the scene is dark in the negative image, and what was dark is light.

NEGATIVE NUMBERS
See Edge Numbers.

NEGATIVE SOUNDTRACK
A soundtrack whose track pattern is black. (See also Positive Soundtrack)

NEGATIVE SPLICE
A 1/16-inch wide overlap splice made on 16mm film.

NEGS
Abbreviation for negatives.

NEPOT (Also Nepotism)
(1) The practice of hiring family members to work on a production or in its offices; (2) the relative who is hired.

NETWORK
An interconnection of affiliated stations bound by contract to an origination point from which television programs and commercials are transmitted via satellite, line feed, or microwave to the local market broadcasting facilities (affiliates). The major U.S. networks are ABC, CBS, NBC, and PBS (a noncommercial network, Public Broadcasting System), and in Canada, CBC. (See also Affiliate, O&O, Networking)

NETWORKING
The interconnection of many broadcast stations for the purpose of simultaneous reception of the same program, used mostly as a cable term. Networking can refer to any joining of stations for the purpose of broadcasting a special program or event; not necessarily only the three major U.S. networks.

NEUTRAL
In color production, black, white, or gray; little or no hue.

NEUTRAL ANGLE
Any camera angle in which the action happens directly in front of the lens. (See also Screen Direction)

NEUTRAL DENSITY FILTER (Also ND Filter)
One of several neutral gray filters or gels that can reduce a light's intensity without changing its color; can be used directly on a camera when the light is too intense for a given shot or f-stop.

NEUTRAL SCREEN DIRECTION
See Screen Direction.

NEVER LEAVE AN AUDIENCE IN THE DARK

A phrase that refers to the practice of always beginning a film before totally turning off the house lights and turning on the lights just before the film ends. The lights are usually dimmed as the first images are projected on the screen and slowly brought back up during the end credits.

NEWS FEATURE

A brief film of seasonal, topical information or human interest shown either during newscasts as a diversion from hard news or on magazine format shows.

NEWSFILM

Film of news items such as interviews, features, or sports items used in television news broadcasts or in news formats, such as "60 Minutes" and "20/20."

NEWTON RINGS

An aberration on the film print that appears in the form of visible ring patterns caused by optical interference when light passes through two film surfaces in imperfect contact in a contact printer.

NG (Also NG Takes)

No good. A term referring to shots that are unacceptable because of technical difficulties, error by the performer, or because of a discretionary decision on the part of the producer or director.

NIELSEN

The A. C. Nielsen Company, one of the major television rating companies, which has developed sophisticated measuring techniques to determine the viewing habits of the American public by conducting an electronic survey of approximately 1200 homes. (See also Nationals, Ratings)

NIGHT FILTER

Used in shooting "day for night"; a filter that reduces the light during exposure, thereby changing the color cast of the shot, resulting in a nighttime effect. (See also Filter)

NIGHT FOR DAY

Shooting a scene during the night and lighting it so that it looks as if it were shot during daylight. (See also Night for Night, Day for Night)

NIGHT FOR NIGHT

Night scenes that are actually shot at night; also the technique for such lighting.

NITRATE

An abbreviated term for cellulose nitrate film. (See also Celluloid)

NITRATE BASE

See Celluloid.

NITROGEN BURST AGITATION

The controlled emission of nitrogen into the film processing bath to induce turbulence.

NODAL POINT MOUNT

A camera mount used when panning glass paintings or background glass to eliminate reflection.

NOISE

Unwanted sound heard or transferred onto a film soundtrack or audio tape from any source, including electronic buzz from the power supply, or atmospheric and environmental sounds.

NOISELESS CAMERA

A camera equipped with a blimp. (See also Blimp)

NONDIRECTIONAL MICROPHONE

A microphone capable of picking up sounds from all directions. (See also Microphone)

NONFICTION FILM

A film that depicts actual events, demonstrates a skill, or presents factual information.

NONSYNCHRONOUS SOUND

See Wild Sound.

NORMAL ANGLE

A shot that takes in the field of action from approximate shoulder height using a 35mm camera with a 50mm (2-inch) lens. (See also Camera Angle)

NORMAL LENS

A lens whose focal length is the diagonal of the film. For example, for 35mm film, a 50mm lens is "normal"; for 16mm film, a 25mm lens, and for 8mm film a 12½mm lens.

NORTH

In animation direction, the upper portion or top of the cel, field chart, or animation table. (See also Animation)

NOTCH

A marking indentation on a film's edge for use in a printer. The notch automatically triggers a change mechanism in the duplicating process, usually involving light exposure intensity, but can be used to effect other modifications in the duplication process.

NOTES

The comments and criticisms by the director and those involved in the creative control of the production of the script or performances.

NUMBERS

Ratings; share of the audience as usually measured by Nielsen or Arbitron.

NUTS-AND-BOLTS FILM

A low-budget educational film, presenting a skill or a how-to theme, usually in a straightforward, unadorned manner. It sometimes refers to a no-frills, often amateurish production.

O

O&O (Owned and Operated)
Broadcast television stations owned and operated by a network or any U.S. corporation. The FCC limits the number of O&Os owned by a network or corporation to seven, allowing no more than five VHF stations and two UHF stations. (See also Network)

OATER
A slang expression for a cowboy (Western) film.

OBJECTIVE
(1) The image forming device of an optical system, usually the lens; (2) the point, or goal, to be achieved in a script or scene.

OBLIGATORY SCENE
A scene that is necessary to take the plot to its expected conclusion; the scene is anticipated by the viewer as it is the logical result of previous action that has led to that point in the plot.

OFF CAMERA
An action, subject, performer, or setting that is beyond or outside of the camera's field of view.

OFF-LINE EDITING
The first stage of tape editing, when creative editing decisions are made. Using a 3/4" videotape cassette, scenes are cut together and a list of the required edits compiled. At the end of an off-line session, a list of all edits (both audio and video) is made for entry into a computer on floppy disc, to be used in the on-line session. (See also On-Line)

OFF-MICROPHONE
(Also Off-Mike)
Any sound that occurs too far away to be picked up by the mike; outside a microphone's particular pick-up pattern. (See also Microphone)

OFF SCREEN (Also OS)
Any action, subject, performer, or prop not seen by the camera, but is presumed to be near the action.

OK TAKES
Shots that have acceptable sound and camera coverage and will be workprinted.

OMNIDIRECTIONAL
(Also Omni)
A microphone with a spherical audio pick-up pattern that makes it equally sensitive to sounds emanating from all directions at one time. (See also Microphone)

ON
(1) A directive that means the performer (or prop, special effect, etc.) is to be on camera immediately; "you're on," is the often-heard

phrase; (2) a switch, circuit, or lamp through which the current is flowing; (3) a slang descriptive used to mean that a performer is energized and ready to perform.

ON A ROLL
Borrowed from gambling, this phrase means that every part of a scene, take, or several scenes or takes in a row, is going smoothly from technical aspects to top-notch performances.

ON-CAMERA NARRATION
A narration during which the narrator is shown on screen.

ON-LINE EDITING
The final stage of tape editing when edits determined in the off-line session are made on a 1" videotape master. At the end of an on-line session, a completed show is available on 1" videotape, from which air masters and dupes are made. (See also Off-Line Editing)

ON LOCATION
Filming, taping, or production that occurs away from the studio or soundstage, whether outdoors or inside another building. (See also Location)

ON SPECULATION (Also On Spec, Shoot on Speculation)
(1) A film produced with independently raised funds, without the backing of one of the major film studios in the hopes of recouping the investment through foreign and/or domestic film rentals, sales, and/or a distribution deal; (2) a script, screenplay, or teleplay that has been written without prior assignment or agreement from a studio, producer, or production company and with the hope of selling it after its completion.

ON THE NOSE
Photography done at the exact exposure indicated by the light meter.

1-LITE
See CRI, Color Reversal Intermediate.

ONE-LIGHT PRINT
A print that is made with the same printer light setting for all shots being printed.

ONE-LINER
A brief retort, usually one line and usually said for a humorous effect, such as a punch line; can also mean literally one line spoken by a performer or indicate a very small role (See also Line); (2) a schedule used in film production that breaks down each day's shooting by one line descriptions, such as "EXT. GAS STATION (D)," and also includes corresponding scene and page numbers, as well as number of pages filmed.

ONE SHEET
Posters used in motion picture theater displays to advertise a film.

ONE SHOT
Frame composition in which only one subject is in the field of vision. (See also Shot)

ONE-WAY SET
A set constructed with only one flat background.

ONES
A term in animation that refers to exposing one frame for each drawing or change of subject. (See also Animation)

OPACITY
The reciprocal factor of light transmission whose logarithm is equal to density (see also Density); the ratio of light upon a film surface and the amount of light transmitted by that surface.

OPAQUE LEADER
See Leader.

OPAQUER (Also Colorer)
A person in animation production who fills in the outlined drawings with solid colors. (See also Animation)

OPENING CREDITS
Titles or credits that appear at the beginning of a film or tape production. (See also Credits)

OPERATING CAMERA OPERATOR (Also Operating Cameraman)
The camera crew member who keeps the camera aimed at the right place during the shot, following the action to the director's specifications.

OPTICAL AXIS
An imaginary line that crosses an optical system's image-forming area to insure symmetry in any plane perpendicular to the line.

OPTICAL EFFECTS
See Effects.

OPTICAL FLOP
(Also Optical Flip)
A visual effect in which the image is flipped over by an optical printer. (See also Flip-over)

OPTICAL HOUSE
A lab or place of business that specializes in optical printing.

OPTICAL NEGATIVE TRACK
See Sound Release Negative.

OPTICAL PRINT
A film print made with an optical printer.

OPTICAL PRINTER
A film printer that consists basically of a camera and a projector. Light first passes through the already developed film that is to be rephotographed, then continues passing through a lens system before it reaches the raw film stock on which the new image is to be printed. The lens system (and related attachments) allows the film to be reduced or enlarged as well as enabling special visual effects to be added at this point. (See also Step Optical Printer)

OPTICAL PRINTING
Printing accomplished by the use of an optical printer as opposed to that done on a contact printer. The image from one piece of film is

transferred to another by light passing through the processed film to the unexposed raw stock. This process allows the duplicate image to be made larger or smaller than the original.

OPTICAL REDUCTION

The process of reducing a larger film format to a smaller one, such as 35mm to 16mm.

OPTICAL SOUND

Sound recorded on an optical soundtrack as opposed to sound recorded on a record (disk), tape, or magnetic film. (See also Optical Soundtrack)

OPTICAL SOUND RECORDER

A sound recorder that produces a photographic soundtrack as distinguished from a sound recorder that records on disk, tapes, or magnetic film.

OPTICAL SOUND RECORDING

A process used today mainly on release prints in which electrical sound signals are converted into a light beam for recording on light-sensitive emulsion.

OPTICAL SOUNDTRACK
(Also Optical Track)

A method of recording sound on film in which the audio signals appear as a photographic image at the edge of a piece of film that, when passed over an exciter lamp beam in a motion picture projector, is translated electronically into a reproduction of the original sounds. (See also Soundtrack, Variable Area Track, Variable Density Soundtrack)

OPTICAL TRANSFER

The transfer of a magnetic soundtrack to an optical soundtrack by using an optical sound recorder.

OPTICAL VIEWFINDER

A range-of-field viewer allowing the director to select appropriate camera angles and setups; it is stereotypically depicted hanging around a director's neck. (See also Viewfinder)

OPTICALS
(Also Optical Effects)

Any of the various effects created to alter a picture visually from the more simple fades used as scene transition to the most complicated special visual effects. The three main ways of creating visual effects and the basic effects most often created by each are as follows: (1) camera—double exposures, split images, creative focusing; (2) printer—fades, dissolves, superimposition; (3) optical bench—wipes, freeze frames, flips. (See also Effects)

OPTICS

(1) The elements that comprise an optical system; (2) the related field of visual effects created by optical systems.

OPTION

(1) A fee paid to hold a person's services or literary property exclusively for an agreed-upon period of time, at the end of which a decision must be made whether to use the services or property or possibly to

extend the option. Options are used to bind performers until a suitable project can be developed or found; (2) an agreement to extend an already existing contract; for example, a producer can require an option in a contract that gives him the right to extend the services of the contractee. Options are usually found in the contracts of today's television performers, writers, story editors, producers, and directors. Although motion picture studios frequently option properties, the option structure is different, because a film is a one-time production opposed to a television series that could run for years. In any case, options are a point of negotiation and are usually handled by lawyers.

ORGAN BACKGROUND MUSIC

Background music played on an organ. Used primarily in the days of silent film, the music set the mood for the audiences then just as film scores do today. (See also Background Music)

ORIENTATION SHOT

An establishing shot that orients the viewer instantly to the locale, mood, or particulars of a scene. (See also Shot)

ORIGINAL

(1) Scripts written specifically for television or film, from the writer's imagination and not an adaptation or based on another work; (2) the first filming, taping, or recording of any film, show, or sound.

ORIGINAL COLOR NEGATIVE

The processed color film as shot in a camera and developed normally to produce a standard color negative in which original colors are represented as their complementary colors.

OSCAR

An award given annually by the Academy of Motion Picture Arts and Sciences for excellence in one of twenty-two categories of motion picture production. The categories are: Best Picture, Actor, Actress, Supporting Actor, Supporting Actress, Direction, Original Screenplay, Screenplay Adaptation, Cinematography, Film Editing, Foreign Language Film, Original Score, Original Song Score or Adaptation Score, Original Song, Art Direction, Costume Design, Sound, Sound Editing, Short Films (Animated and Live Action), and Documentaries (Features and Short Subjects). A winner is selected by secret ballot of voting members belonging to each category. "Best Picture" is the only category for which all members vote, regardless of the category in which they participate. Oscar is the nickname for the statue that, for many years, has been presented on the nationally televised Academy Awards. The statue itself is a gold-plated, bronze, bald, well-formed man. The nickname is said to have been created by an employee of the Academy who, when shown the statue, remarked, "That looks exactly like my Uncle Oscar."

OUT

(1) A script term that directs a specific sound, special effect, or portion of music to be totally stopped, or out, by the specific point indicated; (2) a shortened form of the term out takes. (See also Out Takes)

OUT OF CHARACTER

Exhibiting qualities, displaying physical characteristics, or uttering dialogue contrary to the character's personality as established by the writer or actor. (See also In Character)

OUT-OF-FOCUS DISSOLVE

A transitional special effects technique in which the subject is blurred into nothingness by opening the lens aperture as wide as possible while filming. (See also Special Effects)

OUT OF FRAME

(1) Subjects or action that do not appear in the camera's sight; all that takes place outside the action field; (2) two frames visible on the screen resulting from the projector's frame mechanism being incorrectly set.

OUT OF SYNC

(1) A motion picture sound track that does not correspond with the image or movement on the screen; for example, speech that does not exactly correspond with visible lip movement; (2) the incorrect positioning of the soundtrack on a release print. (See also Synchronization)

OUT TAKES
(Also Outtakes, Trims, Outs)

(1) Shots that, because of technical error, mechanical problems, or mistakes by the performers, are not workprinted. Sometimes, however, because out takes can be quite funny, a group of them are spliced together and shown at, for example, a cast and crew wrap party as entertainment; (2) any shot that is completed and not used in the final edit of the film; (3) any shot that is removed from a film.

OUTLINE

A synopsis of the story elements as envisioned by the writer. This is an early step in the development of a script, before the characters and scenes are fully fleshed out. (See also Script, Story Line, Treatment)

OUTPUT

(1) The connection on a broadcasting, monitoring, or recording unit from which signals go out by means of a cable plugged into the connection through which the signal is fed into other equipment for recording or playback purposes. (See also Input); (2) the power or amount of current available from a source or electrical unit.

OUTRO

The exit speech or tag to a spot or scene. (See also Tag)

OUTWARD THRUST

The direct effect, emotional or physical, that the actions and words of one character have upon another character or upon an element in the plot.

OVER FRAME
A script term indicating that a performer's voice or a sound effect is heard, but the speaker or source of the sound is not visible in the frame of the picture.

OVER MUSIC
Sounds or dialogue heard over a musical background.

OVER SHOT
The filming of more footage than is necessary as a safeguard against technical or other unforeseen problems on critical or difficult footage, or to insure that there will be adequate footage for difficult editing or for a variety of editorial choices.

OVER THE AIR—PAY TV
Television signal that is broadcast via any means such as microwave, or satellite for which there is a fee paid by those receiving the signal. (See also Pay TV)

OVER-THE-SHOULDER SHOT (Also Over Shoulder)
A shot that gives the audience almost the same view as the actor (or subject); the camera is positioned from just behind and to the side of the actor (or subject) and may include a part of the actor's head and shoulder, usually in soft focus. (See also Shot)

OVER-39 LIGHT
See Ambient Light.

OVERCRANK
The operation of a motion picture camera at a speed that is faster than normal in order to produce a slow motion effect of the image being filmed. (See also Slow Motion)

OVERDEVELOP
To darken a film's image either intentionally or by accident by leaving the film longer than necessary in the developing solution.

OVEREXPOSURE
(1) A print that is too light, the result of too much light passing through the lens and onto the film; (2) allowing too much light onto the film or exposing film longer than recommended, usually unintentionally.

OVERHEAD EXPENSES
The ongoing expenses inherent in maintaining a production company, such as insurance, utilities, and rent as opposed to costs derived from a specific production, such as costumes, set construction, and actor's salaries.

OVERHEAD SHOT
A shot accomplished by positioning the camera directly over the action. (See also Craning, Dolly)

OVERLAP DIALOGUE, OVERLAPPED DIALOGUE
See Overlapping.

OVERLAPPING
(Also Matching Action)
(1) Reshooting part of the action of the previous shot or a repetition of action shot, thus avoiding a jump cut in editing; (2) covering the action with two or more cameras and matching the overlapping action in editing, creating a smooth action in

transition and thus also eliminating a jump cut; (3) the instant of superimposure during a dissolve; (4) when referring to dialogue (Overlap Dialogue. Overlapped Dialogue) a line spoken by one actor while another is still speaking his/her line; sometimes done by accident, for example, when one performer who is off camera picks up cues too quickly while the camera is shooting a close-up of another speaking actor; or done intentionally when one actor begins his/her lines while another is still speaking to heighten emotional intensity; (5) when referring to sound (overlapping sound), sound effects, or any audio portion that is carried over into the next shot.

OVERLOAD

(1) Recording sound at a level higher than a system can accommodate, which results in sound distortion; (2) the use of too much current for the fuses or cable to handle.

OVERNIGHTS

A service provided by a television ratings company that evaluates the numbers of viewers of shows on an overnight basis. Surveys are made from data gathered in the major markets (Los Angeles, Chicago, Philadelphia, Detroit, San Francisco, New York); the remainder of the approximately 150 markets in the nation are surveyed, and the results are made available in two or three days after the programs air.

OVERPOWER

The technique of adding ample light of the correct color temperature to compensate for the effect of existing fluorescent lights; it is usually necessary to add more light than would be used in nonfluorescent lighting conditions.

OVERRIDE

See Manual Override.

OVERS

(1) Unnecessary footage of shots that are too long to be used; (2) film deliberately exposed in the camera in order to use up remaining film.

OXIDATION

The weakening of certain photochemicals used in developing film due to the accidental presence of oxygen.

P.A.
(1) Abbreviation for production assistant or program assistant; (2) public address system.

PACING (Also Pace)
The tempo or rhythm of a story or live production. The speed with which the plot unfolds. Also, the balance of slower, dramatic scenes with comedic relief or musical numbers with monologues.

PADDLE PLUG
A flat plug used on stages for supplying electrical current to various pieces of equipment used in production.

PAINT FOREMAN
The person in charge of painting sets.

PAINTING ON FILM
Using transparent or opaque paints to apply effects or images directly onto clear or translucent film, done either frame by frame, or by applying desired colors over entire sections of film. (See also Cel, Special Effects)

PAN (Also Panning)
(1) Pivoting the camera horizontally to capture action, subject, or background, sometimes used to mean "follow the action"; (2) filming a shot at one angle and moving the camera to another angle without interruption in the movement or stopping the camera; (3) any pivotal movement of the camera. (See also Sky Pan, Swish, Tilt, Vertical Pan)

PAN AND TILT HEAD
(Also Pan-tilt head)
A camera mount that allows for the smooth vertical and horizontal rotation of the camera; the mount may be attached to a tripod, dolly, or clamp.

PAN FOCUS
A large depth of field that allows the entire field of action to be in focus. (See also Focus)

PAN HANDLE
A handle attached to a pan-tilt head to add needed leverage for smooth movement of the head.

PAN SHOT
(1) A shot filmed while the camera pivots on its tripod in a horizontal arc; (2) any shot that uses panning. (See also Pan, 360° Pan)

PANAVISION
A complete system of cameras and optics, available through lease only from Panavision, Inc. This system or any of its components cannot be purchased, which enables the company to maintain consistent

PANCAKE

quality at the highest state of the art.

PANCAKE

A commonly used, heavy coverage makeup base available in all shades ranging from white to natural skin tones, as well as colors such as green, blue, and red for special effects or clown makeup. (See also Makeup)

PANCHROMATIC FILM

Black-and-white film sensitive to the entire visible spectrum that when exposed creates a negative image. This film must be contact printed on panchromatic film or paper to produce a positive image.

PANIC LIGHTS

See Emergency Lights.

PANORAMA

An all-encompassing extreme wide-angle view of a scene revealed in a pan shot or presented as a single shot.

PANTOGRAPH

A hanging support of variable length for stage lights.

PANTOMIME

Performing without words, using only movement and gestures to convey a character, action, or story.

PAPER EDIT

See Dry Edit.

PAPER PRINT

A motion picture print made on a nonacetate base, but one whose support is paper.

PARABOLIC SPOTLIGHT

A light that utilizes a parabolic reflector to obtain and project a narrow beam of light.

PARALLACTIC MOVEMENT

What appears to be the movement of two objects past each other, which in reality is the movement of the camera.

PAR LAMP

See Birdseye.

PARALLAX

The difference or displacement in the field of action, especially noticeable in close-ups, between the way an image appears in a viewfinder and the way it appears as seen through the lens of a camera. When taking a shot of a building from the street, the walls of the building will converge toward each other at the top. Independent or monitoring viewfinders are usually mounted above or to the side of a camera lens, and parallax must be taken into consideration. (See also Viewfinder)

PARALLAX CORRECTION

Any method used to eliminate compositional errors due to parallax. (See also Parallax)

PARALLAX ERROR

The result of not correcting for viewfinder parallax. (See also Parallax)

PARALLEL

A portable scaffolding system with a narrow platform that accommodates a small crew; from it lights are suspended. Cameras can also be attached to accommodate the camera crew for high-angle shots.

PARALLEL ACTION
Two or more themes running concurrently through a story. The intent is to convey to the viewer that the action is occurring simultaneously and is shown by cutting back and forth from one story to the other.

PARODY
A comedic imitation of a great or well-known work.

PART
A role or character in a script that will be brought to life by an actor. (See also Role)

PATCH
To join one piece of equipment to another by using a patch cable whose plugs fit into connections in patch panels found on consoles and audio and video components.

PATCH CORD
A short cable used in patching to connect one piece of equipment to another.

PATCH PANEL
The originating point at which cables leading to components are connected.

PATCHING
The process of connecting a patch. (See also Patch)

PAY TV
A generic term used to include the various subscriber television services, including cable TV systems, that present, usually for a monthly fee, uncut, first-run films, sports programs, educational courses, and other such programming that is usually unavailable otherwise. (See also Over the Air)

PAY YOUR DUES
A colloquial expression meaning an individual must be willing to work in the industry in a subordinate position for as long as it takes to make the contacts needed to get ahead.

PEDESTAL
A strong metal support, usually on wheels for easy mobility, used to hold a light or camera.

PEG BOARD
(1) A board used in production and editing containing a series of numbered or labeled pegs on which pieces of workprint film are stored by scene number for easy access in making a shot breakdown or editing composition; (2) in animation and title work, the surface that holds the cards or cels steady under the camera by pins stuck through holes in the margins of the artwork. (See also Animation)

PEGGING
The swing of the needle on a VU meter into the red zone and all the way to the right, caused by a sudden burst of loud sound or the master sound volume being set too high.

PELLICLE
A sheet of mirrored film.

PENCIL TEST
An animation technique in which rough penciled drawings on white paper are filmed and projected at normal speed in order to test the speed and smoothness of cartoon movement. (See also Animation)

PENETRATION
The number of subscribers to a cable television system in ratio to the number of households, or potential subscribers, in the cable system's area. Penetration largely determines a system's profitability. (See also Cable)

PER DIEM
A Latin term that loosely translated means "daily." In the industry sense, it is additional money allowed per day for living expenses while on location. Per diem is given to each member of the cast or crew separately from their regular paycheck.

PERAMBULATOR
The mobile platform that holds the boom microphone and boom operator.

PERF
The commonly used abbreviation of the word perforation.

PERFORATIONS
(Also Sprocket Holes)
Holes manufactured in exact predetermined spacing along one or both sides of motion picture film that are used to position and move the film frame by frame through a camera, projector, or printer. Each film gauge, or size, such as 16mm or 35mm, has its own specified perforation spacing.

PERKS
The free extras that might come along with the business, such as free tickets, satin jackets with a studio logo, free travel with the production company, fancy parties, premieres, and many other large and small privileges.

PERSISTENCE OF VISION
The visual phenomenon in which the eye retains an image for a moment after it is seen. This allows for the individual frames of a film when projected to be seen as a continuous picture without flicker.

PERSONALITY TEST
An on-camera interview in which performers are asked simple questions, usually about themselves, in order to relax them while enabling their personality to come across on the screen. (See also Audition)

PERSPECTIVE DISTORTION
An appearance of distortion in photographic images when the subject is seen from a point other than the center of perspective or when it is viewed with a wide-angle or a telephoto lens.

PERSPECTIVE DRAWING
A drawing that illustrates the design and dimensions of the upright sections of a set, giving the illusion of depth. (See also Set)

PHANTOM
A ghost image within another image. (See also Ghost Image, Halation)

PHASING
(1) The hollow, swooshing sound

resulting when two microphones connected to a single amplifier are poorly placed; (2) a sound distortion effect, sometimes created deliberately as a special effect, when two identical tracks are played into the same amplifying source slightly off synchronization, with one track slightly faster or slower. The resulting sound is a rushing, spacelike effect.

PHI PHENOMENON

The visual phenomenon in which two objects appear to change positions instantly, without movement.

PHOTO-CUT-OUT GLASS SHOT

A glass shot that is made by shooting through a piece of glass to which a portion of a photograph has been attached as a matte to provide additional scenery.

PHOTO MATTE

A portion of a photograph that is used as a matte. (See also Matte)

PHOTOELECTRIC CELL

An electronic device that varies in response to the intensity of the light striking it, also called an electric eye. In projectors, variations in light caused by the sound track are converted by the photocell into current that can be used as an audio signal.

PHOTOFLOOD

An incandescent lamp manufactured to produce more than normal light output and color temperature when used at normal voltage; these bulbs have a shorter life but are used commonly in motion picture production. Loosely, any bulb with such characteristics.

PHOTOMETER

A less commonly used term for "light meter."

PHOTOMETRY

The science of the measurement of light.

PHOTOPLAY

See Screenplay.

PHYSICAL TIME

(Also Production Time)

The actual length of a production in hours and/or minutes as opposed to dramatic time in a script.

PICK IT UP

(1) An order, usually from the director, to continue the scene from where it was stopped, without changing the camera angle; (2) a directive to the performers to speed up the action or intensify or brighten the energy of a scene.

PICK-UP

(1) Going back to reshoot a scene or portion of a scene after the main filming or taping is completed. For example, after shooting a show before a live audience, the cast and crew usually stay to do pick-ups of certain lines or parts of the production that could be perfected; (2) an option agreement written into the contract of a performer or key creative person (such as a writer) that, if picked-up, secures the person's services for a specified length of time; (3) pick-up of a show by one of the major networks, meaning that the

show will be broadcast for a specific number of episodes or airings on that network. (See also Pick-Up Deals)

PICK-UP DEALS

Negotiated points in some contracts to retain the services of any cast or crew members (including the writer, director, or producer) or product use (network airing of a show or series) for a specified period of time at a specified amount of money. (See also Option, Pick-Up)

PICK-UP RECORDER

A magnetic-sound recorder with the capability of erasing tracks or portions of the mix and rerecording over them without noticeable clicks at the start and stop points.

PICTURE (Also Pix)

(1) A motion picture, usually a feature film; (2) the isolated video or image portion of a film or tape as opposed to the film or tape complete with soundtrack; (3) A still photograph. (See also Still)

PICTURE DUPLICATE NEGATIVE (Also Dupe)

(1) A negative printed directly from a master positive. This negative is used to make other positive copies of a film; (2) a reversal printed directly from the picture negative.

PICTURE MASTER POSITIVE

An intermediate special print made in producing a picture duplicate negative.

PICTURE PRINT

A film print with positive images.

PICTURE RELEASE NEGATIVE

(Also Release Negative)

An edited negative that will be used to print release prints. (See also Release Prints)

PIEZOELECTRIC MICROPHONE

See Crystal Microphone.

PIGTAIL

A device consisting of a three-wire hook-up used to tap into 230-volt lines to power lights for production.

PILOT

A television show (the first episode in a series) created and produced in the hope it and the series will be purchased by a network. Frequently pilots are aired on a network on a one-time basis when the network has decided not to buy the series. (See also Backdoor Pilot)

PILOT PIN

See Registration Pin.

PILOT PIN REGISTRATION

A system that exactly positions successive frames of motion picture film in perfect relationship to the film's sprocket holes with precisely sized and controlled pins that hold the film still and in position in a camera, printer, or projector. (See also Registration)

PILOT PRINT

In order to reduce costs, a first trial print in black and white is made from a color original. By using various colored filters during projection, the production staff will be able to tell how the footage will look when printed in color.

PIN SCREEN ANIMATION

An early animation technique, developed in 1932 by Alexandre Alexeieff, in which thousands of tiny pin holes are made in a metal sheet; varying the height of the pins forms images. (See also Animation)

PINHOLE

A clear tiny circle that appears on the developed film emulsion, usually caused by inadequate agitation of the processing fluids.

PINK NOISE

A staticlike sound created in all audible frequency ranges in order to test a theater's sound system.

PIPE GRID

See Grid.

PIRATED PRINT

An illegally duplicated copy of a film, tape, or recording. (See also Bootleg)

PISTOL GRIP

A handle that allows for a more secure hand grip on a camera or mike when attached to the bottom of a hand-held camera or microphone.

PITCH

(1) To submit a story idea, plot line, or project for consideration by a production company. Well-known producers and writers often can sell a project by pitching a story, as opposed to making a formal presentation; (2) of perforation pitch, the distance between the leading edges of two sprocket holes on motion picture film, pitch is smaller for original films than for print films or duplicates; (3) the relative musical tone position of a sound by which it can be judged to be high or low, sharp or flat.

PIX

Abbreviation for pictures.

PIXILATION ANIMATION
(Also Pixilation)

An animation technique using live performers to achieve an animation effect of people moving by having the performers move slightly between exposures and holding each new position while one or two frames are exposed. (See also Animation)

PLAN

A drawing done to scale of the top view of a set. (See also Set)

PLANE OF CRITICAL FOCUS

The area of a picture that is the clearest and sharpest in terms of focus. (See also Focus)

PLANO-CONVEX SPOTLIGHT

A spotlight that produces a narrow beam of light using a plano-convex lens.

PLANT
The deliberate conveyance of information to the audience in order to prepare them for some future event in the story.

PLATE (Also Process Plate, Key)
A photograph on a sheet glass support used as background when front or rear projection is being used. (See also Process Projection)

PLATEN
A flat piece of glass used to press animation material on an animation stand. (See also Animation)

PLAYBACK
(1) The procedure of checking sound or video just recorded for quality; (2) any machine that reproduces previously recorded audio or video material (3) a prerecorded audio tape and the process of playing it back during filming when it is more efficient or economical to use a prerecorded track than live sound.

PLAYBACK TRACK
A specially prepared soundtrack to be used during filming as a performer's cue or background sound during the action.

PLOT
The series of events and character relationships and actions that are the foundation of a story.

PLOT GIMMICK
(Also Gimmick, Plot Plant, Boff, Weenie)
An idea of action, mechanics, effect, or any element that is created to move the plot along.

PLUG
(1) To promote a film, television show, or performer in the media; a slang expression for publicity or advertisement. Many celebrities appear on talk shows to plug their new film because this can help sell tickets; (2) an electrical connector.

PLUG IN
To connect electrical components.

PLUGGING BOX
(Also Spider Box)
A device made up of several female electrical plugs to which lights and other electrical equipment can be connected.

PLUS LENS
See Close-up Lens.

POCKET
A permanently mounted studio electrical outlet, usually one with protective covers.

POINT OF VIEW (Also POV)
A shot made from the line of sight of the performer, with the camera showing the audience the action from the viewpoint of the actor.

POLAR DIAGRAM
A head on, point-of-view diagram of the pickup pattern capability of a microphone.

POINT OF ATTACK
The selected point in a story at which the writer begins the script.

POLARIZED FILTER
A filter with optical slits that can be lined up with the plane of the

polarized light, to pass polarized light, but which can reduce polarized light when the slits are turned at right angles to the plane. (See also Filter)

POLARIZED LENS
A lens with an attached polarizing filter. (See also Lens)

POLARIZED LIGHT
Light vibrating in one plane only, as commonly found in reflections from most surfaces, except for shiny metals such as chrome.

POLAROID FADE
A fade accomplished with a pair of properly placed (depending upon type of fade desired) polaroid filters in which one filter rotates at a 90° angle to the other.

POLAROID FILTER
The trade name for a commonly used polarizing material that when used in front of a camera lens will effectively eliminate unwanted polarized reflections from a scene. (See also Filter)

POLAROID SHOT
A Polaroid picture taken at the end of a shot, used to assist the director and camera operators to position performers and props exactly to avoid a jump cut and to keep continuity.

POLE RIGHTS
Rights that allow a cable television system to install coaxial cable on existing utility poles by agreement with a telephone or power company. (See also Cable)

POLECAT
Metal tubing used to form scaffolding to support lights or other rigging, or tubing that can be edged between walls, the floor, and ceiling to support various pieces of equipment.

POLISH
To rewrite or perfect a line, a scene, or an act of a script. Sometimes a writer other than the original one is called in to polish or punch up a script.

POLYESTER FILM BASE
(Also Polyester Tape Base)
A tape and film base composed of synthetic resin.

POOL HALL LIGHTING
A lighting technique that uses a single light source, usually hanging in the middle of the set, which can be seen on camera.

POP FILTER
A filter that when placed on a microphone will eliminate popping and breath blasts sometimes present in sound recordings, live audio situations, and in sound systems.

POP IN (Also Pop On)
An animation technique in which additional art indicators such as arrows or lines can be placed between single-frame exposures, directly on the copy to be photographed, which results in their sudden appearance at that point in the frame sequence.

POPPING
An undesirable sound caused by

emphasis placed on consonants, especially the letter "p" in speech, which is sometimes present on sound recordings and can be heard through a faulty microphone and amplification system.

PORNOGRAPHIC FILM
(Also Blue Movie, Porno Film; Porno Movie; Porno Flick)

A commercial film whose subject matter is explicit sex. (See also Skin Flick, Soft Porn Film)

POSITIVE (Also Positive Film)

(1) A photo image that has the same tonal values as those of the subject, a natural print as opposed to a negative image, whose tones are reversed; (2) a film with positive images. (See also Negative Film)

POSITIVE PITCH

The various perforation differences found among standard print film.

POSITIVE SCRATCH

The black image scratch intentionally placed on film that appears on the positive from which the print was made. (See also Scratch, Scratched Print)

POSITIVE SOUNDTRACK

A soundtrack, used on release prints, whose track pattern is clear as opposed to the black pattern found on negative soundtracks. (See also Negative Soundtrack, Soundtrack)

POSITIVE SPLICE

An overlap cement splice having the exact width of the film being spliced, for example, for 16mm film, a splice having a width of one-eighth-inch. (See also Splice)

POSTFLASH (Also Postfog)

A technique of reducing the contrast of a photoimage by exposing the already exposed, but undeveloped, film to a low level of light, thereby causing a slight overall fogging.

POSTPRODUCTION

All work done after filming or taping is completed, including editing, looping, printing, laying in music, and sweetening.

POSTPRODUCTION SERVICES

The services involved in the completion of a film after it has been shot and workprinted, including printing, editing, looping, and mixing. (See also Postproduction)

POSTRECORDING

Recording sound or dialogue after filming or taping has been completed. (See also Postsynchronized Sound)

POSTSCORING

Recording music to suit a film that is already edited.

POSTSYNCHRONIZED SOUND

Any sound effect or lip-synched dialogue that is added after the film or tape has been shot. (See also Dialogue Replacement)

POV

Abbreviation for Point of View.

POWDERMAN
The colloquial name for the crew member who is responsible for designing and executing explosions and fires for the special effects department.

POWER BELT
A battery belt.

POWER CABLE
Any cable used to connect various pieces of electrical equipment to a power source.

POWER PACK
A battery set used to supply power to cameras and recorders. The pack may be a separate unit, worn on a belt, or attached to the camera itself.

POWER ZOOM
A motor drive that powers a zoom lens.

PRACTICAL PROP
A prop that is operational, i.e., a lamp, a stove, or a door, as opposed to a prop that is for appearance only. (See also Action Prop, Decorative Properties, Prop)

PRATFALL
A sight gag in which a performer falls and lands on his behind. (See also Gag)

PREBREAK
To partially break a breakaway prop so that when used it will break more easily; for example, in a fight scene if a broomstick is to be broken over a stuntperson's head, the broomstick will be prebroken.

PREEMPT
To cancel temporarily or reschedule a television program in order to put in its place another show, sporting event, or important news broadcast.

PREFLASH (Also Prefog)
A technique of reducing the contrast of a photo image and increasing its sensitivity to light by exposing the film to a small amount of light before it is used in a camera.

PREMIERE
(1) The first public showing of a film; (2) opening night of a production complete with press, hoopla, and a party after the screening.

PREMIX
The combining of several soundtracks onto one track, which eventually will be combined with other tracks. Premix is usually done when there are more tracks to be combined than there are playbacks.

PREPRODUCTION
All duties that are performed prior to shooting, including scripting, casting, and location scouting.

PREQUEL
A film set in a time period earlier than that of a film already released, using many of the same characters. (See also Sequel)

PRERECORDED
Recorded in advance of airing; especially in live shows, such as awards shows and beauty pageants that are prerecorded from a live

show on the East Coast for later broadcast in other time zones.

PRESCORING
Recording a sound track that will be used later during shooting. A prescored track is often used when filming musical numbers in undesirable locations or under noisy conditions. (See also Playback)

PRESENCE
The quality of recorded sound that gives the audience the impression that the source of that sound is in the room in which the action is occurring. Auditory cues such as room noise and tonal quality are used to convey presence. (See also Room Tone)

PRESET CONTROL
A device programmed to activate lights at a predetermined level of intensity.

PRESS AGENT
A colloquial term for publicist, not necessarily a member of the Publicist's Guild. Colloquially speaking, one who plans promotional ideas for the purpose of securing free publicity through the media. The press agent is responsible for the liaison between the media and the celebrity, film, or show being promoted.

PRESSURE PLATE
A mechanism in a camera, projector, or printer that presses the film against the aperture plate during exposure or screening.

PREVIEW
(1) Screening of a feature film before it is released to the public; (2) short excerpts of a motion picture shown in theaters before the film is released as a form of advertisement to create the desire in the audience to see the film when it is released. (See also Trailer)

PREVIEW PRINT
(Also Sample Print)
A copy of a feature film provided by the studio, producer, or distributor to prospective customers or to critics, prior to general release.

PRIMARY COLORS
In the spectrum of light, blue, red, and green are the primary colors. In all cinematography, lighting, filtering, and color processing, these are the colors from which all hues are mixed. The primary colors are manipulated by using their complementaries; yellow, magenta, and cyan. (See also Color Separation, Complementary Colors)

PRIME ACCESS
The 30-minute time period that just precedes Prime Time, usually 7:30 P.M. until 8:00 P.M., as set aside by the FCC for airing of local programming as opposed to programming supplied by a network, with the exception of interrupting for national coverage of a sporting event or a major national news story such as an election. (See also Prime Time)

PRIME LENS
A lens that has a fixed focal length.

PRIME TIME

The television viewing period as defined by the FCC between 8:00 P.M. and 11:00 P.M. in the Eastern and Pacific Time Zones, and from 7:00 P.M. until 10:00 P.M. in the Central Time Zone. In the Mountain Time Zone, each station selects whether prime time shall be 6:00 P.M. to 11:00 P.M. or from 5:00 P.M. until 10:00 P.M. Cable stations use 6:00 P.M. until 11:00 P.M. The largest audiences watch television during these hours. (See also Daytime, Early Fringe Time, Family Hour, Fringe Time, Late Fringe Time, Prime Access)

PRINCIPAL PERFORMER

A lead actor in a continuing role or starring role.

PRINCIPAL PHOTOGRAPHY

The filming or taping that commences after preproduction, the main photography of a film with the performers but excluding editing or pick-ups. When principal photography is completed, postproduction begins.

PRINT

(1) The positive copy of a piece of film; (2) "Print it"—the order given by the director when a take is satisfactory and is to be workprinted. The camera operator notes it on the script and log (See also "Save It," "Take It"); (3) an additional copy made from the master; (4) advertising or promotion of a production that appears in newspapers or magazines.

PRINT FILM (Also Print Stock, Print Film Stock)

Film manufactured to be shot, developed, and projected with positive images and soundtracks.

PRINT PITCH

The standardized distance between perforations of manufactured print film stock as opposed to the shorter pitch distance of film manufactured for use in cameras.

PRINT REVERSAL

See Reversal Print.

PRINT STOCK

See Print Film.

PRINT SYNC (Also Printer's Sync, Projection Sync)

See Projection Synchronization.

PRINTER

A machine that transfers the images from one strip of film to another, usually from exposed and processed film onto raw stock. There are several different kinds of printers including continuous, step, contact, and optical. The printers work by passing two films (or more on sophisticated printers) in front of an aperture with regulated light.

PRINTER FADERS

(1) A device on a printer that makes fades; (2) a fade-out or fade-in made by a mechanism on a printer that gradually decreases or increases the light on a predetermined number of frames. Usually fades are made from a positive; but negative film can be used, if the

light techniques are reversed, that is, to fade-out, increase the light and to fade-in, decrease the light. (See also Fade)

PRINTER HEAD

The mechanism on a printer in which the printer light is located and through which the two films being used come together.

PRINTER LIGHT

A light source in a printer. Its color temperature is carefully regulated.

PRINTER LIGHT SCALE

A scale of printing-light intensity that allows original film images to be printed darker or brighter as desired and to produce an evenly exposed print from an original with uneven exposure levels. The printer light scale is also used in the execution of special optical effects such as dissolves and fades.

PRINTING

The process of duplicating a film.

PRINTING ROLLS

A picture or soundtrack roll processed to run through a printer. (See also A&B Rolls, Optical Sound)

PRINT-THROUGH

An unwanted effect that can occur on tightly wound magnetic audio tape in which the signals leak through the coating of adjacent layers of the tape, causing an echo effect when the tape is played; usually caused by high recording levels or too high a temperature where the tape is stored.

PRINTING THROUGH THE BASE

The technique of reversing the print film or print stock so that the light from the printer penetrates the base layer prior to the emulsion. This technique is sometimes used to make a B-wind print from a B-wind original, for example.

PRINTING WIND
(Also A-Wind)

The direction in which film prints are turned so that when the prints have been rolled through the printer and their sprocket holes are in alignment, the emulsions of the two films will be facing each other in optical printers, or in contact with each other in contact printers.

PRISM SHUTTER
(Also Prism Intermittent)

A rotating prism of four or more sides through which viewer light passes as film is pulled through, a device used on many film viewers and editing machines.

PROCESS BODY

Part of a car body or other vehicles used in a studio in front of a process screen. The vehicle appears to move against the background on the screen, when in fact it is stationary.

PROCESS CAMERA

A camera specifically designed for special effects work, that is, matte photography and bipack printing, as opposed to other cameras such as studio or field cameras, which are used for filming live action.

PROCESS CINEMATOGRAPHY
Cinematography that uses the process camera.

PROCESS PHOTOGRAPHY
Shooting background scenes that will be used in rear projection work in a studio.

PROCESS PLATE
A background image on a positive lantern slide that will be used for rear projection. (See also Plate)

PROCESS PROJECTION
The technique of projecting a background, using either front or rear projection, onto a translucent screen, in front of which live action is photographed. (See also Process Projector)

PROCESS PROJECTOR
A projector interlocked in sync with cameras so that their shutters open and close simultaneously. This results in the projection of a series of backgrounds that move across a translucent screen, in front of which actors are photographed while performing.

PROCESS SHOT
A shot accomplished by photographing live action in front of a screen upon which background images are projected. (See also Process Projection)

PROCESSING
The developing and printing of motion picture film, including all the physical and chemical steps necessary to convert exposed film stock into a satisfactory picture.

PRODUCER
A person, persons, or company who make feature films, television programs, or any theatrical venture. The producer finds or creates a project and oversees its production, from raising the funds to making the deal for its showing. The producer has the ultimate responsibility for the shaping and end result of a production.

PRODUCTION
(1) All of the processes that go into the making of a feature film or television show, i.e., writing the script, setting the budget, scouting location, casting, and principal photography; (2) the film or show itself.

PRODUCTION ACCOUNTANT
One who keeps track of the outgoing expenses and arranges for payment of services and production equipment associated with the shoot.

PRODUCTION ASSISTANT
A nonmanagement position on the crew in which one distributes copies of script changes, arranges for rehearsals, prepares and distributes rehearsal call sheets, times each scene, calculates overall timing, and assists in the day-to-day requirements of production. Production as-

PRODUCTION BOARDS

sistants actually assist the director, not the producer.

PRODUCTION BOARDS
See Boards.

PRODUCTION BOOK
(Also Production Digest, Board Breakdown)
A detailed record documenting updated information regarding all aspects of production including the script breakdown, transportation requirements, costumes required, locations to be used, and budget information. The production book is the source of information used when laying out the boards. (See also Boards)

PRODUCTION CAMERA
A live-action camera, such as a field camera or a studio camera. Not the same as a process camera, which is used to achieve optical and special effects, or an animation camera, used solely for filming cels.

PRODUCTION CODE OFFICE
(Also Code Office)
The offices and organization through which the Motion Picture Association of America enforces its Production Code.

PRODUCTION COORDINATOR
One who assists in general day-to-day duties of production, such as meal arrangements and distributing schedules. Unlike the production assistant who assists the director, the production coordinator deals with peripheral duties of production, not with what actually happens during shooting.

PRODUCTION CREDITS
A listing of the names of all who worked on the production. The list appears at either the beginning or the end of a production. (See also Crawl, Credits)

PRODUCTION DIGEST
See Production Book

PRODUCTION MANAGER
The individual whose responsibility it is to oversee and manage the business affairs of a production, such as cost controls, the arranging and scheduling of all production elements (see Production Schedule, Production Report), and cast and crew logistics. This position reports directly to the producer and may hire unit managers and assistants, in accordance with need and union rules.

PRODUCTION NUMBER
(1) A part of a film or show that requires a great deal of production, staging, or effects, above and beyond what is needed for the rest of the production, especially song and dance numbers; (2) an identification number used to facilitate allocation of money and the assigning of costs and related accounting information.

PRODUCTION OVERHEAD
All costs relating to the production of films or television shows accrued by studios and production companies that are not in direct re-

lationship to one specific project, for example, general administrative costs, salaries of production department executives, secretaries, and staff, project abandonment costs, studio or office rental costs, and general operational costs.

PRODUCTION PERSONNEL

Individuals involved in the production of a film or television show. A list of the most common positions includes Art Director, Assistant Director, Choreographer, Continuity Person, Head Gaffer, Best Boy, Casting Director, Director, Director of Photography, Editor, Gaffer, Grip, Juicer, Makeup Artist, Mixer, Producer, Production Supervisor, Script Person, Second-Unit Director, Special Effects Supervisor, Stunt Supervisor, Technical Advisor, Unit Manager, and Writer.

PRODUCTION REPORT
(Also Daily Production Report)

A daily report issued by the production manager outlining in detail the crew and performers who worked on a particular day, the hours worked, the amount of footage shot on that day, as well as the cumulative figures and other information pertinent to the production that indicates the film's current financial and time-frame situation.

PRODUCTION SCHEDULE
(Also Shooting Schedule)

A schedule handled by the production manager that lists the sequence of the shots to be made, along with a listing of the crew, performers, equipment, and transportation needed for each shot.

PRODUCTION SCRIPT
See As-Broadcast Script.

PRODUCTION SECRETARY

A secretary to a member of the production staff.

PRODUCTION TIME

Another term meaning physical time, as opposed to dramatic time. (See also Physical Time)

PROGRAM SEPARATOR
(Also Bumper)

A five-second announcement between a show and a commercial; for example, "We'll return after these messages."

PROGRESSION

The sequence of events as they occur after the exposition, leading to the climax of the piece as the forces in conflict move toward one another.

PROJECTION

(1) Speaking in a volume loud enough to be heard by the audience; (2) throwing an enlarged image onto a screen.

PROJECTION BOOTH

A small room, usually located in the rear of the theater, in which projectors are operated.

PROJECTION-CONTRAST ORIGINAL

An original reversal film designed for normal contrast, shown through a projector.

PROJECTION LEADER
A leader, used specifically in projectors, that enables the projectionist to make an instant changeover from one film to another. (See also Academy Leader, Leader)

PROJECTION SPEED
The rate at which film is run through a projector, i.e., 24 frames per second for sound films; 18 frames per second for silent films. (See also Silent Speed)

PROJECTION SYNCHRONIZATION
(Also Projection Sync)
The alignment of picture and sound on a print in such a way that they are in sync for a particular type of projector. For example, there is a 26-frame displacement between the sound and the picture on 16mm film. (See also Sound Advance)

PROJECTIONIST
The individual who operates and maintains projectors.

PROJECTOR (Also Film Projector, Motion Picture Projector)
An electronic device in which motion picture film is threaded and passed over a light beam, which projects the images on the film onto a screen; projectors can also be equipped to reproduce sound from a filmed soundtrack.

PROJECTOR SPEED
See Camera Speed, Frame Fate.

PROMO
Short for promotional announcement.

PROMOTIONAL ANNOUNCEMENT
(Also Promo)
A brief advertisement to promote a television show, series, or feature film.

PROMOTIONAL FILM
A film, produced usually by a corporation or a business, used for publicity purposes.

PROMPTER
An automatic cueing device that displays a script in rough synchronization with the action, especially used by television newscasters.

PROP
An object used on the set during taping or filming to augment a script. Any portable object used on a set is considered a prop. Props are either hand props, such as a gun, book or dish; action props, such as a ball or bicycle; or set props, such as table lamps, books in a bookcase, or pictures on a wall. The formal term is "property" which is rarely used in day-to-day conversation. (See also Action Prop, Decorative Properties, Hand Properties, Practical)

PROP MAN (Also Prop Person)
A crew member whose responsibility it is to obtain and maintain properties used throughout the filming.

PROP PHRASES
In dialogue, particular phrases

that convey important information to the audience.

PROP SET
A small set that only suggests a location or environment with symbolic details, as opposed to a large, intricate set that creates or duplicates a location. (See also Set)

PROPERTY
(1) A literary work, such as a book or a short story, whose film rights are owned or are the subject of negotiations by a producer or production company; (2) a formal term for prop (See also Prop), seldom spoken but used more often in contract language.

PROPERTY HANDLER
(Also Prop Handler)
One who makes certain the props required for each scene are on the set in proper position and condition for use.

PROPERTY MASTER (Also Prop Master, Prop Supervisor)
The person in charge of obtaining and maintaining all props required in production.

PROPERTY SHEET
A listing of props that are needed during production.

PROPERTY TRUCK (Also Prop Truck)
The truck or vehicle used for transporting props from the storage area to the set or location.

PROPMAKER
The individual who makes props that will be used for a specific purpose during production.

PROSPECTUS
See Bible.

PROTECTION COPY
A copy made from the master tape of a television show and delivered along with the air print to the network as a backup copy.

PROTECTION MASTER
A duplicate of a film or tape from which duplicate copies are made, instead of using the original print, which might be damaged during reproduction.

PROTECTION SHOT
A duplicate shot made in order to give the editor insurance against poor continuity; done especially with scenes that would be expensive or impossible to reshoot. (See also Shot)

PROXIMITY EFFECT
An effect accomplished by a cardioid microphone, whereby lower frequency sounds are accentuated when the sound source is close to the microphone. (See also Cardioid)

PROXY
A character written into a script for the sole purpose of imparting information to the audience.

PROXY QUESTIONS
A writing technique used to convey information to a viewer by the dialogue triangle, that is, by having one character ask a question of an-

other (See also Dialogue Triangle, Proxy)

PUBLIC ACCESS

A requirement that the general public be allowed to present programming on cable television. The FCC requires that major cable markets provide for local use, one channel for educational use, one channel for local government, and one channel for public access programming. The channels reserved for educational and local government use must be available free of charge for at least five years. It is not required that production facilities be made available for educational and local government programming. The free public access channels must be available indefinitely, and public access channels are required to make available at least the minimum production facilities necessary for public access programming.

PUBLICITY STILL PHOTOGRAPH
(Also Publicity Still)

A still photograph taken during production, or sometimes a still printed from a frame of film, used to advertise or promote a film.

PULL BACK

To move the camera back from the action, to dolly out from a close-up.

PULL FOCUS

(1) The act of changing the focus during a shot; (2) to follow focus. (See also Focus, Follow Focus, Rack Focus)

PULL UP (Also Tighten)

(1) To reframe a shot to eliminate extraneous background; (2) to eliminate unnecessary material from a script.

PULLDOWN

The act of moving the film frame-by-frame through the camera or projector using the pulldown claw. (See also Intermittent Movement)

PULLING

The process of identifying and separating the original film that will be used from the film that will not be used in the final version of the film. (See also Matching, Original)

PUNCH

(1) In a creative sense, to punctuate dialogue with a joke or to pick up the general tone, such as to punch up the second act where it drags; (2) in audio or videotape production, to cover a mistake by literally punching in the play button at the pick-up point and recording over it. In recording music, a singer's lines can be punched in and resung at any place in a recording to attain desired perfection in a song; (3) in synchronization, a device used to punch a hole in film leader to locate a starting point for editing or printing synchronization; (4) in animation or title work, the device that punches holes in the edges of animation cels or title cards to accommodate the pegs on an animation stand or easel (See also Animation); (5) in blooping, to punch a specially shaped hole in the soundtrack area of a film negative to reduce the sound of a bloop.

PUNCH UP
(1) The directive to punch a line or effect into a track; (2) to enliven a performance or dialogue.

PUP LIGHT
See Baby.

PUPPET ANIMATION
A type of animation using puppets with exchangeable heads and body parts, with slightly different expressions and positions. (See also Animation)

PUSH
To force process film. Certain films can be shot at a faster speed than specified by the manufacturer and be pushed in the processing lab to accommodate the faster speed.

PUSH OFF
(Also Push-over Wipe)
A wipe effect in which one image pushes another off the screen. (See also Wipe)

PUSHING
See Forcing.

PUT A HANDLE ON IT
Add plausible dialogue or action to tighten up loose points in a story line.

PVC BASE
A base for tape or film made of polyvinyl chloride.

PYROTECHNICS
Fireworks used as special effects in television, production numbers, and films.

PYROTECHNIST
One who creates and executes fireworks used as special effects in television and film.

QUANTEL
A digital video effects unit that manipulates a videotape signal, thereby allowing the operator to create a variety of visual effects, including enlarging or reducing the image and making the image spin.

QUARTER-INCH TAPE
(Also Audio Tape, Magnetic Tape)
Standard width magnetic audio recording tape used on reel-to-reel recorders and in tape cartridges for general audio and soundtrack purposes.

QUARTZ LAMP
(Also Tungsten-halogen Lamp)
A bulb manufactured with a halogen, thereby lengthening its filament life and allowing the filament to better maintain color temperature as the bulb does not darken and change color toward the end of its life. (See also Halogen Lamp)

QUARTZ LIGHTS
A luminaire which uses a quartz or tungsten-halogen bulb. (See also Halogen Lamp, Quartz Lamp)

QUICK AND DIRTY
An informal term which refers to a rough print of a film needed right away; with no concern for quality except that the picture and sound be in sync.

QUICK CUT
(1) A very brief take or short piece of film, to be inserted into a shot or scene during editing; (2) a written directive in a script that indicates that the action is briefly to be interrupted by a reaction or another action.

R

R LAMP
See Birdseye.

RACK
(1) A frame used for mounting electronic equipment for storage or use on the set; (2) a frame of rollers that carries film through processing; (3) to focus a lens.

RACK FOCUS (Also called Pull Focus, Shift Focus)
To change lens focus while shooting.

RACK-FOCUS SHOT
A shot using rack focus to change the focus and depth of field and thus to emphasize another part of the action. (See also Focus, Follow Focus, Shot)

RACKOVER
(1) A camera mechanism that allows a viewfinder to be located behind the lens, where the film is normally located, thus permitting viewing of the scene directly through the lens; (2) to move the rackover into position.

RACKOVER VIEWFINDER
An internal viewfinder, located behind the lens in the spot normally occupied by the film, that allows the camera operator to view the scene directly through the lens without parallax. (See also Rackover)

RADIO MICROPHONE
(Also Wireless Microphone)
A microphone equipped with a miniature transmitter and an aerial that allows for operation within a limited range without a cable connecting it to the power source. (See also Microphone)

RANGEFINDER
An instrument through which the distance to the subject from the camera is determined by bringing together the split or double image viewed through the rangefinder. This type of device is used in certain kinds of aerial shots, predominantly helicopter shots, but rarely for other types of cinematography.

RANKING
The ratings number from 1 to 64 that tells what position a television show is in, according to audience surveys made weekly by Arbitron and Nielsen.

RATING
(1) A letter code (G, PG, R, and X) administered by the MPAA and given to feature films to indicate audience suitability (See also MPAA Code); (2) An estimate of how many households are viewing a particular television show at a given time period. (See also Arbitron, Nationals, Nielsen)

RATING SERVICE

An independent organization whose function it is to survey the markets for listening and viewing habits. The Neilsen Company and Arbitron are the two major rating services in the country.

RAW STOCK

Unexposed, undeveloped, and unprocessed film that is ready for use in a camera or printer; also unused magnetic tape or film.

REACTION SHOT

A shot that shows an actor's reaction or emotional response to dialogue directed at him or her or to action going on or that has just occurred. (See also Cutaway, Shot)

READ-OUT LINES

The lines on lenses and light meters that lead from one scale to another, for example, the line from the depth of field scale to the footage scale.

READ THROUGH
(Also Reading, Table Reading)

The first step in rehearsal. The actors, producers, writers, and director read the script for the first time as a group.

READER (Also Sound Reader)

A mechanism used in film editing to pick up and play back magnetic or optical sound through small speakers or headphones.

READING

(1) Auditioning for a part by reading a scene selected by the casting director or producer; (2) the earliest stage of rehearsal (See also Read Through); (3) the measurement on light or VU meters and referred to as light or sound reading.

REALISM

Using scripts, locations, costuming, performances, and camerawork to create and present action as real, as opposed to fantasy.

REAR PROJECTION
(RP, Rear Screen Projection, Back Projection)

Projection of backgrounds onto a translucent screen in front of which actors are filmed. When it can be used, rear projection eliminates the cost of building a set or of location shooting.

REAR-PROJECTION UNIT

A projector whose motor is interlocked with that of the camera. While the projector is projecting a backdrop on a rear projection screen, the camera is photographing the action that is taking place in front of the screen.

RECANS

Pieces of raw stock cut from long pieces or left over from a shoot. (See also Short Ends)

RECORD FILM

A noncommercial film usually used for analytical purposes, such as movies of sporting events in order to study the team plays or home movies for family enjoyment.

RECORDER

A device that records sound, picture, or both on magnetic audiotape or videotape. Also, any device used to record sound on magnetic film, motion picture film, or disc.

RECORDING
Preserving sound for reproduction. (See also Recorder)

RECORDING LEVEL
The input signal volume level as set by the gain control and registered on a VU meter.

RECORDING STUDIO
An acoustically perfected room or studio used for recording music, sound tracks, wild tracks, and looping, usually consisting of separate units of two rooms each. One room or portion of a room is used to record the music or performer and the other room is the control booth with the console, recorder, and producer.

RECORDIST
The individual who operates recording equipment.

RED LIGHT
A rotating red light outside a stage door, sometimes accompanied by a bell or buzzer, that alerts everyone that filming is in progress. When the light is on, absolute silence must be maintained and no entry or exit is permitted until the light goes off.

REDRESSING
The act of replacing furniture and props on a set exactly as they appeared at the beginning of the shot. This is done when additional shooting, such as retakes of a fight scene, or pick-ups to correct errors in the actors' performances, is required.

REDUCTION DUPLICATE
The print that results when a larger film stock is reduced to a smaller stock. (See also Reduction Printing)

REDUCTION PRINTING
The process of duplicating a film onto narrower film stock, such as going from 70mm to 35mm or from 35mm to 16mm.

REEL
(1) A spool made of either metal, wood, or plastic, around which lengths of film are wrapped; (2) ten minutes of premeasured sound film.

REEL CAPACITY
The amount of film that can be wound on a reel. The standard capacitities are: 25, 50, 100, 200, 400, 600, 800, 1,200, 1,600, and 2,000 feet per reel.

REEL-TO-REEL
A winding technique of film or tape in which reels instead of cassettes or endless loops are used to feed and take up the film or tape; also, the type of recording or playback equipment that uses reels.

REEL-TO-REEL TAPE RECORDER
A tape recorder that uses feed and take-up reels instead of cassettes or cartridges. (See also Reel-to-Reel)

REESTABLISHING SHOT
A shot that appears in the middle or end of a scene in order to show

once again the general location in which the action takes place. Sometimes it is made from an angle different from that used in the establishing shot. (See also Establishing Shot, Master Shot, Shot)

REEXPOSURE (Also Flashing)

A second exposure given to reversal positive film after it has been processed, usually to add special effects or correct color or light.

REFLECTED LIGHT METER

A light meter that registers the amount of light reflected from the subject.

REFLECTION

The bouncing off or throwing back of light or image from a reflecting surface.

REFLECTOR

(1) Any object, usually a type of foil, with a light reflecting surface used to redirect light to fill in shadows; (2) any surface used to redirect sound waves.

REFLECTOR LAMP

A luminaire with a built-in reflector.

REFLEX CAMERA

A camera that allows the operator to view the field of action through the lens while it is filming, by intercepting and reflecting the light passing through the lens to a viewfinder, thus eliminating parallax.

REFLEX FOCUSING

The capability to focus the camera by looking through the lens, which has a reflex system, a feature found on most newer cameras. The older models require the lens to be relocated to a different position before reflex focusing can be accomplished.

REFLEX FOCUSING DEVICE (Critical Focusing Device, Critical Focuser)

A device that allows the camera operator to obtain exact focus. (See also Focus, Reflex Focusing)

REFLEX SHUTTER

A shutter that reflects light passing through the lens into the viewfinder by way of a mirror set at 45° to the lens axis.

REFLEX VIEWFINDER

A viewfinder that has no parallax and allows for right-side up images to be viewed through a camera lens. (See also Viewfinder, Zoom Finder)

REFRACTION

The bending of a ray or wave of light as it passes from one medium to another, such as through a glass of water.

REGISTER

(1) To indicate emotion, i.e., a performer registers pain to react to a stimulus; (2) to hold film precisely in place as it is exposed; (3) accurate superimposition of two or more images in any frame of registration printing.

REGISTRATION

(1) The ability of a camera or projector to align film, frame by frame, and to hold it steady while it is being

exposed or projected; (2) the precise positioning of film in the aperture gate of a camera, projector, or printer in exactly the same place for each consecutive frame; (3) when referring to animation, the exact positioning of layers of cels. (See also Registration Pin)

REGISTRATION PEGS

Pegs on platens and animation tables that are identically aligned so that when artwork is prepared on platens and moved to the animation table, the framing will match.

REGISTRATION PIN

A pin that enters the film perforation during the exposure in a camera and holds the film in a fixed position.

REGISTRATION SHOT

A shot made from a still photograph or diagram in which the first few frames on the film are in precise alignment with the still (See also Shot)

REGULAR REFLECTION

Reflection that occurs naturally from bright, smooth surfaces and for which lighting compensation must be made.

REGULAR 8 (Also Cine 8, Standard 8)

Eight millimeter film with forty frames per foot, one perforation per frame line and with perforations the same size as those on sixteen-millimeter film.

REHEARSAL HALL RUN-THROUGH

A rehearsal that takes place in a rehearsal hall as opposed to a run-through that is held in a studio using cameras and recorders. (See also Run-through, Walk-through)

RELATIONAL EDITING

A visual metaphor that places two images in juxtaposition in order to compare or contrast their relationship. For example, a shot of a large flock of sheep and a shot of a crowded New York sidewalk.

RELATIVE MOTION

The movement of one object in relationship to another. Sometimes in film work it is easier to replace real motion of the subject with relative motion of the camera. Sometimes an object that seems to move on the screen, such as a train, is actually filmed with the train immobile but with the camera moving at an appropriate speed and angle, giving the illusion of the train's motion.

RELEASE

The initial distribution of a film for exhibition in commercial theaters; (2) shortened term for press release, a purportedly factual, usually promotional, informational news sheet regarding a film, TV show, performer, industry event, or studio news release; (3) short for release form.

RELEASE FORM

A form that, when signed by a performer or writer, allows a producer to use his or her material, i.e., dialogue, photograph, or film clip without the usual contracts and payment.

RELEASE NEGATIVE
A duplicate negative from which additional release prints are made. (See also Picture Release Negative, Release Print)

RELEASE PRINT
The final, edited composite print, after the approval of the final trial composite print, which is ready for general distribution; the film that is seen by the public.

REMAKE
An updated or new version of an old film.

REMBRANDT LIGHTING
See Back Light

REM-JET BACKING
See Antihalation Backing.

RENTAL
The use of a film for a specified length of time in exchange for a fee paid to the distributor.

REPRISE SHOTS
Shots that are reshown at some later point during a film.

RERECORDING
The transfer of sound from one medium to another, or from one track to another, i.e., transferring sound effects, music, dialogue, from separate tracks to one track. (See also Mixing)

RERELEASE
The release of a film for showing in commercial houses after a length of time has elapsed since its previous release.

RERUN
The showing of a movie or television show after it has once appeared on the air.

RESIDUALS
Money earned from television programs that are rerun after their first showing. (See Rerelease, Rerun) The residual scale is set by the union to which the payee belongs, i.e., AFTRA or the Writers Guild; however, it can be negotiated to a higher price in special circumstances. An average performer's residual payment structure is approximately half of the original salary for the first rerun, 25 percent for the second rerun, 15 percent for the third rerun, and a very small amount for every other rerun thereafter.

RESOLUTION
Action that occurs after the climax. (See also Denouement)

RESOLVER
The device in a tape playback machine or magnetic film recorder that controls the speed of the sound in relationship to the picture.

RESOLVING
The process of transferring a quarter-inch, lip-synced soundtrack tape to magnetic film. The control signal recorded on the quarter-inch tape is used as a guide to synchronization during transfer. (See also Control Signal)

RESPONSE CURVE
The curve that represents the ability of sound to maintain a specific frequency at a certain volume.

RETAKE
The immediate reshooting of identical action; a retake filmed or taped later is a pick-up. (See also Pick-up)

RETICLE LINES
Guidelines in a camera viewfinder that indicate specific points of reference such as the center of the frame and the television safe-action area.

RETICULATION
A defect in which the surface of the film emulsion cracks and wrinkles in a shattered, ridged pattern that ruins the film; caused by too high a temperature in film processing, especially radical temperature change when the film is wet. When this occurs the film cannot be salvaged.

RETROFOCAL LENS (Also Inverted Telephoto Lens)
A special lens designed to increase optically the physical size of the subject and that has an unusually convenient, easy-to-mount-and-operate system. (See also Lens)

REVERBERATION
An echo; multiple or repeated reflections of sound whether created intentionally or through technical error.

REVERBERATION TIME
The amount of time necessary for a sound to be reduced to one-millionth of its initial intensity.

REVERSAL ORIGINAL
A reversal film manufactured for exposure in the camera.

REVERSAL POSITIVE PROCESS (Also Reversal Process)
The development technique by which positive film, either exposed in the camera or printed from a positive, is developed as a positive. The stages for this process are: first development, bleaching of the developed image, reexposure, second development, fixation, washing, and drying.

REVERSAL PRINT (Also Reversal Master Print)
A film duplication technique in which a direct transfer can be made from a positive to another positive using special reversal film, thereby eliminating the need for an intermediate negative.

REVERSE ANGLE
A shot from the opposite (or reverse) POV of the preceding shot; a shot made from a 180° change from the preceding shot. For example, filming two people seated at opposite ends of a table; the camera faces the first person and films, then reverses the angle and films the second person. The camera usually would set up where the first person was sitting for the reverse angle shot. (See also Shot)

REVIEW
A critical evaluation of a film or television show. (See also Critics)

REWIND (Also Rewinder)
An electric or manual mechanism used for rewinding film on a projector or reel-to-reel.

REWRITES
See Change Pages.

RHEOSTAT
Proper term for a dimmer that varies voltage control and allows for lights to be slowly dimmed or brightened, as desired. (See also Dimmer)

RIDE FOCUS
The focusing ring adjustment made on a lens to keep the action in perfect focus during a shot. This procedure may include automatic parallax correction of the viewfinder on some cameras.

RIDE GAIN
(Also Ride the Pot)
The act of controlling the amplitude of incoming or outgoing audio signals.

RIFLE SPOTLIGHT
(Also Rifle Spot)
A type of spotlight that is elongated and ellipsoidal.

RIG (Also Rigging)
(1) Ropes, blocks, and equipment associated with a fly system; (2) the act of setting the rigging.

RIGHT-READING
(1) Graphic material positioned so that letters are in natural reading order from left to right, as opposed to wrong-reading in which letters are backwards (See also Emulsion Position, Wrong-reading); (2) proper lateral position of images, such as a tree in normal position with its branches in the air, as opposed to having its roots in the air. (See also Lateral Orientation)

RISER
See Apple Box.

ROADSHOW
A selected release of a feature film only in major theaters and at higher admission prices.

ROCK VIDEO, ROCK TV
A music video format; or a video of a rock song. (See also Music Video)

ROLE
(1) A part in a film; (2) a character portrayed by an actor in a film. (See also Part)

"ROLL"
(1) A directive to start cameras and recorders; (2) a spool, reel, or core-load of film; (3) camera rotation around the lens axis.

"ROLL CAMERA"
See "Roll 'em."

"ROLL 'EM"
(Also "Roll Camera")
A command given by the stage manager at the director's instruction to begin cameras and recorders.

ROLLER BANK
See Spool Bank.

ROLLERS
A free turning device around

which film is threaded and wound, used to control the film as it winds through equipment.

ROLL FILM MOTION PICTURE CAMERA
A term describing a motion picture camera that uses roll film as opposed to film cartridges.

ROLL SOUND
A command to start the sound recorders.

ROLLOFF
The gradual weakening of sound.

ROLLUP
An electronic mechanism through which rollup titles are displayed, by moving them upward at rates that can be changed.

ROLL-UP TITLES
Titles prepared for use in a roll-up, arranged vertically on a flexible background material.

ROOM TONE (Also Room Noise, Room Sound, Presence)
The background sounds or noise present even in the most quiet setting. This ambient noise is undesirable if too loud, however, low levels create realistic sounds and are often recorded and used to fill in the silent passages on a sound track, or to back up dialogue and narration tracks that may have been recorded without room tone.

ROTARY MOVEMENT
A special whirling image effect created during printing by use of an optical spin attachment on an optical printer.

ROUGH CUT
The first edited version of a film. The shots and scenes are laid out in sequence. A more detailed editing job is the next step. (See also Workprint)

ROYALTY
A fee payable to an author or publisher for the right to use copyrighted material often payable each time a work is performed or broadcast.

RUMBLE POT
A device used to obtain a fog effect in which crushed dry ice is placed in the rumble pot, which is full of boiling water.

RUN LINES
A method of rehearsing by having someone read dialogue of the other characters while the actor reads his or her own lines.

RUN-THROUGH
A rehearsal, usually done with cameras and sound recorders monitoring the action but not actually recording.

RUNAWAY PRODUCTION
A term applied to those productions done outside of the Hollywood community or in a foreign country in order to economize costs.

RUNNING SHOT
A shot accomplished by a camera operator running with a hand-held camera. (See also Shot)

RUSHES
See Dailies.

S

SAFE-ACTION AREA (Also Television Safe-Action Area)
The area within a camera frame (outlined in the camera's viewfinder) that will be visible in over-the-air television transmission; the areas outside of the frame will be automatically cropped in transmission.

SAFELIGHT
A low-wattage colored light bulb, usually red, used in darkrooms enabling workers to see and handle film during processing without damaging the film. The film is not sensitive to deep red.

SAFETY BASE
(Also Safe Base)
A slow-burning film base, most commonly cellulose triacetate, as distinguished from the highly flammable nitrate film bases used in the early days of film. (See also Acetate Base, Nitrate)

SAFETY FILM
Film whose emulsion is coated onto a safety base; any film that uses a safety base, as do all commonly used modern film stocks. (See also Safety Base)

SALAD
A colloquial term that refers to the pile-up of film after it has jammed in a camera, printer, or projection unit. (See also Jam)

SAMPLE PRINT
See Final Trial Composite Print, Preview Print.

SANDBAG
A bag of sand; often used in production as anchors to hold in place scenery, certain set frames, and light stands.

SATURATION
The degree of intensity and purity of color. A color is saturated when it is at its clearest and most brilliant. It is least saturated when the color is washed out and faded. An oversaturated color appears dense and muddy.

"SAVE IT,"
(Also "Take It," "Print It")
An order from the director that means the take is to be workprinted.

"SAVE THE LIGHTS"
(Also "Save the Arcs")
A directive to turn off the lights on the set.

"SAVE THE FOOD"
A colloquial command to fake a particular action during rehearsal and wait for the take to do it.

SC
Abbreviation for Scene.

SCALE
See Minimums.

SCANNING
(1) Colloquially, to fast forward a soundtrack or videotape with the audio on cue or video moving quickly but discernably on a monitor in order to locate a specific area of the track or tape; (2) a television optical sound process in which a narrow light beam or electromagnetic radiation strikes all parts of a modulated area (optical sound recording) or on one which becomes modulated in the scanning process.

SCENARIO
(1) A script or screenplay; (2) a plot outline, or an outline of future events; (3) a setting. (See also Scene)

SCENARIST
A dramatic script writer.

SCENE
(1) A group of shots that together forms a complete unit of action, a succession of one or more shots within a sequence; (2) various actions not necessarily restricted to one place at one specific time that, when dramatically linked, comprise an act in a television script; (3) descriptive of a shot, such as a love scene, a crowd scene, or a war scene; (4) a setting, such as, the scene is a park in winter or Paris in the spring.

SCENE CONTRAST
In the action field, the degree to which tones change quickly from light to dark.

SCENERY
(1) Flats or any artificial backgrounds used during shooting, such as sides of buildings or skylines; (2) natural objects found on location and used as background while shooting.

SCENERY DOCK
A storage and loading area for scenery and flats.

SCHEDULE (Also SKED)
An itemized breakdown of television programming; or, a sequence of events in a production, such as the rehearsal schedule, or the shooting schedule.

SCHUFFTAN PROCESS
A visual special effect created by the camera photographing a subject or background reflected in a mirror, and simultaneously filming action seen through an area of the mirror where the mirror silver has been removed.

SCIENCE FICTION FILM (Also Sci Fi Film)
A genre of film that combines scientific fact with fantasy to create plausible, or somewhat plausible, plots, universes, theologies, environments, and living beings, usually spiced with many special effects, often, but not necessarily, with an outer space theme. *The Day the Earth Stood Still* is a science fiction classic from 1951. Modern classic examples include *Close Encounters of the Third Kind* and the *Star Wars* trilogy.

SCIOPTICON
A projection machine that uses moving slides.

SCOOP
A 500-1,500 watt floodlight with a wide, diffuse, soft, round, and largely uncontrollable beam.

SCORE
(1) The act of composing and/or matching music to a film; (2) the music itself, a soundtrack.

SCOUTING LOCATIONS
(Also Scouting)

The search for appropriate locations for each shot; to look for the proper locale for an entire production; usually done by the director or a location scout.

SCRATCH
To mark a film abrasively, either intentionally or by accident; or such a mark that appears on the film.

SCRATCH-OFF ANIMATION
Using artwork painted with water-soluble paint and photographed upside down, animation is created by erasing small portions of the artwork between exposures. The film is then processed and turned rightside up, returning the film to its proper position for viewing. (See also Animation, Shooting Upside Down)

SCRATCH PRINT
(Also Slash Print)

A rough print used only as a guide in postproduction dubbing, mixing, music, or narration recording, and similar situations where the visual quality is of little importance. Scratch prints are usually untimed black-and-white prints made from assembled originals (A & B rolls), without printer-light corrections. A scratch print is not physically marred or scratched as is a "scratched print." (See also Scratched Print)

SCRATCH TRACK
A sound recording of lesser quality used to determine exact dialogue in postrecording or postproduction. (See also Guide Track)

SCRATCHED PRINT
A motion picture print that has been intentionally rendered useless by making a heavy scratch down the length of the film. This is a technique sometimes used when a demonstration copy of a film is needed to eliminate the risk of possible duplication or pirating of the print.

SCREEN
(1) A flat, reflective surface on which motion pictures, slides, or backgrounds are projected; (2) to project a film; to view a film, as a screening.

SCREEN BRIGHTNESS
The luminance reflected from a screen, as measured in candles per square foot.

SCREEN CREDITS
(1) Titles listing the names and jobs of those people involved in production of a specific motion picture (See also Billing, Credits, Titles); (2) a chronology or list of all of the films in which an individual has been involved, whether he be director, producer, performer, or crew member of any kind.

SCREEN DIRECTION

The direction in which the action appears to move across the screen, often expressed in compass points, such as, east-west to indicate right-to-left movement, north-south to indicate top-to-bottom movement. Screen direction is determined by the position of the camera relative to an imaginary line, or axis, drawn through the dominant action in each shot. The three main categories of screen direction are: (1) constant screen direction, in which the movement does not change from shot to shot; (2) contrasting screen direction, in which the movement does change from shot to shot; and (3) neutral screen direction, in which the movement is toward or away from the camera.

SCREEN EXTRAS GUILD
See Extras.

SCREEN RATIO
See Aspect Ratio.

SCREEN TEST

A short test film of one or more scenes, made to determine whether a performer is the best choice for a particular role. Sometimes many actors who are in the final casting sessions for a part are asked to screen test to narrow down the finalists. (See also Audition)

SCREENING

As most commonly used, a preview showing of a feature film. Screenings are often held for press critics and a select group of the target audience to either gain an honest reaction (See also Sneak Preview) or to generate positive feelings about the film.

SCREENING ROOM

A room in which films are screened for noncommercial purposes.

SCREENPLAY

A script written specifically for a motion picture. In television production, the script is referred to as a teleplay. (See also Script, Teleplay)

SCREENWRITER

A writer of stories, treatments, and scripts for motion pictures.

SCREWBALL COMEDY

An American film genre especially popular in the 1930s, in which likeable characters from clashing social backgrounds were depicted in comical, fast-moving, impossible, satirical, and incongruent situations, mostly involving sex and romance.

SCRIBE

A sharp, pointed metal tool or stylus used in editing to mark the film edges by scratching in the emulsion information needed by the editor for conforming.

SCRIM

A thin, gauzelike material, either mounted on a frame and placed in front of a light source, or placed directly over a light, to reduce the intensity of sunlight if shooting outdoors, or diffuse and soften any artificial light. (See also Dot, Finger, Flag, Gobo, Target)

SCRIPT
The written pages of a story broken down into acts and scenes, as conceived by the writer, and presented in a typed format. A script includes dialogue, stage direction, descriptions of the locales and settings, suggested camera angles, narration, music, special effects, and any other details required by the writer, director, and producer. A script follows a basic transition from outline and rough draft to the final shooting script, whether it is written for television or feature films. Even film shorts, cartoons, documentaries, and news programs require a script. (See also Screenplay, Teleplay)

SCRIPT BREAKDOWN
The breakdown and arrangement of shots so that all scenes set in the same location can be shot at one time. Although this requires shooting out of sequence, shooting the scenes in this manner is cost effective. (See also Boards, Breakdown)

SCRIPT PERSON (Also Continuity Person, Script Girl)
The production person whose key job it is to assure that all details in one shot, including costumes, positioning, and props, remain consistent through all similar shots and takes. This person also records information important to the director and the editor.

SCRUB
A colloquial term meaning to delete from a shooting schedule.

SEASON
Usually twenty-four consecutive episodes of a television series; the new television season begins in the fall, and continues for approximately twenty-four weeks. The networks contract with the producers to purchase the desired number of episodes to be aired that season. If a show is cancelled midseason, a replacement is scheduled by the network to air in that time slot. (See also Mid-Season Replacement)

SECOND-GENERATION DUPLICATE
(Also Second-Generation Dupe)
A film or tape copy made from the first generation film or tape. The first generation is not the original but is the first duplicate made from the original. Each duplication is called a generation; with each generation the quality of the dupes slightly degenerates. (See also Generation)

SECOND UNIT
An additional film crew consisting of a second unit director, a small camera, and sound crew. Second units often get fill shots or shots that complete a scene, such as an airplane taking off, waterfalls, other scenery, and small bits of action.

SECONDARY COLORS
Same as Complementary Colors.

SEGUE
The act of achieving a smooth transition from one piece of business to the next. For example, in segueing from one piece of music to another, the first piece fades as the next one immediately begins. (See also Cross Fade, Sound Dissolve)

SELECTIVE FOCUS
The technique of using shallow depth of field to bring into sharp focus only a portion of the action field. (See also Focus)

SELF-BLIMPED
A term that describes a camera specifically designed to produce such little noise that a separate blimp is unnecessary.

SELF-MATTING PROCESS
See Dunning-Pomeroy Self-Matting Process.

SELSYN (Also Selsyn Motor)
A self-synchronizing electric motor used with many synchronizing projection, sound recording, and camera systems.

SENIOR
A focusable studio lamp consisting of a Fresnel lens and a 5,000-watt bulb.

SENSITIVITY
The degree to which film emulsion responds to light.

SEQUEL
A film or television production that follows a previous production, usually beginning where the previous one ended and using some, if not all, of the same characters. *Rocky* is a popular film with two sequels. (See also Prequel)

SEQUENCE
A series of shots or scenes connected by continuous theme and purpose.

SEQUENCE SHOT
A long shot, in which the actors and camera can move about; this eliminates the delay involved with separate setups for medium and close-up shots.

SERIAL
A series of episodes that are released separately. Each episode ends by leaving one of its heroes or heroines in jeopardy or at a dramatic peak. At the beginning of each following episode, the action of the preceding show is recounted to bring the audience up to date. Soap operas are considered serials.

SERIES
A group of individual television episodes with the same basic characters in each installment, but whose plot is dramatically complete in each episode. "All in the Family," "M*A*S*H," and "The Waltons" are examples of primetime series.

SERVO MOTOR
A self-regulating speed control motor system for cameras and sound recorders.

SET
Short for setting, although set has become the term in general usage. A location that has been constructed and assembled on a stage in which the action is photographed; any location where filming or taping takes place. (See also Four-Wall Set, Three-Wall Set, Two-Wall Set, Plan, Prop Set, Unit Set, Wild Wall)

SET DRESSER
The crew member who is responsible for placing the furnishings on a set.

SET DRESSING
(1) Furniture, draperies, and related objects used on the set; (2) the process of placing the furnishings on a set.

SET LIGHT
See Base Light.

SET PROP
See Prop.

SETUP, SET UP
(1) Setup: the plan and arrangement of the cameras and lights in order to shoot the action from all angles and positions for the best effects, as created by the director; (2) Set up: the actual setting up of the cameras and lights for action; (3) dialogue or situations created by one or more characters in order to set up the scene for another character's joke or reaction, or for a believable plot sequence, or to impart certain information to the viewer. (See also Breakdown)

70MM FILM
A relatively modern film size producing large screen results; an original camera film that is often reduced to 35mm stock (reduction duplicate) for showing on theater screens unable to accommodate 70mm.

SEXPLOITATION FILM
A film with poor production values that uses sex as its subject in order to attract an audience. (See also Exploitation Film)

SFX
Abbreviation for Sound Effects.

SHALLOW DEPTH OF FIELD
(Also Shallow Depth, Shallow Focus)
A limited area of clarity and definition in the field of action that noticeably leaves other areas out of focus. (See also Focus, Selective Focus)

SHARE
A term referring to the percentage of the total population watching a specific television program at a particular time. (See also Arbitron, Nielsen, Rating)

SHARP
(1) Having clarity, definition, and detail easily visible; (2) a lens able to record detail well.

SHOOT
(1) A colloquial term meaning "to film" or "to photograph," (2) the actual filming or taping of a production; also a generic term used to refer to filming or taping a production.

SHOOT AROUND (YOU)
A term that refers to a rearrangement of the shooting schedule in order to accomodate a performer who is temporarily unable to perform. This is not a procedure that is taken lightly, and is only used when a performer is absolutely unable to meet his call due to illness or a personal emergency of some kind.

SHOOT ON SPECULATION
To film without a deal or contract in hopes of selling the production when completed. (See also On Speculation, Speculation Film)

SHOOTING
All the related activities involved in a shoot or actual production.

SHOOTING CALL
See Call.

SHOOTING LOG
See Camera Log.

SHOOTING RATIO
The ratio of the amount of film shot to the length of the edited film. For instance, if a production shot 20,000 feet and the edited film was 5,000 feet in length, it would have a 4 to 1 ratio.

SHOOTING SCHEDULE
(Also Production Schedule)
A day-by-day breakdown of the scenes to be shot, the cast needed in those scenes, the props, the wardrobe, and the production personnel required, and all particulars needed for a specific day's shoot.

SHOOTING SCRIPT
The final approved script that is followed by the director and performers during production.

SHOOTING TO PLAYBACK
A filming technique whereby the performers lip sync to dialogue or music that has been prerecorded. Filming with a prerecorded track allows for fewer problems than with live sound recording and gives the performers more freedom of movement during filming.

SHOOTING UPSIDE DOWN
In order to make the action appear to go in reverse, the camera is upside down while shooting. The film is processed, and turned right side up for projection. Double perforation stock should be used for this. (See also Scratch-Off Animation)

SHOOTING WILD
Filming without sound.

SHOPPED
Descriptive of a property that has been pitched and rejected by many or all of the networks, pay TV, or cable companies.

SHORT ENDS
That portion of unused raw film stock near the end of a roll that is too short to be used. (See also Recans)

SHORT FOCAL LENGTH LENS
A lens with a wider-than-normal field of view that gives a feeling of depth to the set; a lens from 35mm to 9mm. (See also Focal Length, Normal Lens, Wide-Angle Lens)

SHORT SUBJECT (Also Short)
A brief film usually shown before the main feature in a commercial theater. Cartoons and travelogues can be considered shorts; however, shorts are also live action on a variety of subjects. (See also Single Concept Film)

SHOT (Also Take)
A single piece of action photographed in an uninterrupted flow. Various types of commonly used shots include: Aerial, Angle, Close-up (CU), Crane, Cutaway, Down, Elevator, Dolly, Establishing, Extreme Close-Up (ECU, XCU), Ex-

SHOT BREAKDOWN

treme High Angle, Extreme Long Shot, Extreme Low Angle, Eye Level, False Reverse, Follow, Full, Glass, Grease-glass, Group, High-Angle, Incoming, Insert, Library, Long (L.S.), Loose, Low-Angle, Master, Matte, Medium (MS), Medium Close (MC), Medium Long (ML), Mirror, Moving, One, Orientation, Over-Shoulder, Protection, Rack Focus, Reaction, Reestablishing, Registration, Reverse Angle, Running, Skull, Swish Pan, Tank, Tilt, Three, Three-quarter, Tight, Traveling, Trucking, Two, Up, Zoom. (See individual designations for complete definitions.)

SHOT BREAKDOWN

(1) A listing of the shots to be made in sequence, along with the names of the performers, crew, and equipment required; (2) the sequential filming of action from different camera angles or positions.

SHOT LIST

A list of those shots that contain neither dialogue nor narration.

SHOT PLOT

The detailed guide of camera angles as they correspond to shots described in the script, drawn on a set or location plan; the width of the angles listed indicates the focal length of the camera lens at which a scene will be shot.

SHOTGUN MICROPHONE
See Cardioid.

SHOULDER BRACE

A portable framework that supports a camera and can be worn by a camera operator to permit mobility while shooting. (See also Camera Mount)

SHOULDER POD
See Shoulder Brace.

SHUTTER

A mechanism that opens and closes the lens aperture of a camera. The device alternately shuts as unexposed film is pulled into the gate and opens for the specified amount of time indicated by the aperture setting, which allows light to pass through the lens and the film to be exposed. Also, a similar device in a projector that controls the length of time that light, passing through the film, remains on the screen by cutting the projection light during the time the film is moving at the aperture. (See also Shutter Speed, Variable Opening Shutter)

SHUTTER ANGLE

The size of a movie camera's shutter opening measured in degrees. Since the opening is wedge-shaped, it is measured in angular degrees.

SHUTTER SPEED

The length of time a camera shutter is open, allowing exposure of the film, usually measured in fractions of a second.

SIBILANCE

An undesirable hissing sound or dominance of "s" sounds in speech over a microphone.

SIDE BACKLIGHT
See Kicker Light.

SIDES

Pages of dialogue, excerpted from a script or specially written, selected by the producer and/or casting director and given to actors auditioning for a specific part. (See also Audition, Camera Test, Test)

SIGHT GAG

A joke created visually, without dialogue. (See also Gag)

SIGHT LINE

The view or line of vision from any seat in a theater to the screen or stage.

SIGNAL-TO-NOISE RATIO

The noise present in a sound system when there is no other input.

SILENT FILM

(1) A film without sound; (2) a film that was made before 1928 in the days of silent films. Most theaters, however, provided some kind of live music played in the background.

SILENT PRINT

A film print without sound or soundtrack.

SILENT SPEED

The speed at which silent film was shot and should be projected; normally 18 frames per second. In the early days of silent films, the speed was 16 frames per second, although projectionists were often told to crank faster for chases and more slowly for love scenes. (See also Projection Speed)

SILHOUETTE ANIMATION

The animation of a silhouetted figure. (See also Animation)

SILVER HALIDE

A light-sensitive silver compound used in film emulsions, such as silver chloride, silver bromide, silver iodide, or silver fluoride.

SIMULTANEOUS CONTRAST

A visual phenomenon in which hues or color values appear to change, depending upon their background. For example, medium gray appears light gray against a dark gray background, but appears dark gray against a light background; magenta looks more red against a blue background and more blue when placed against a reddish background.

SINGLE

(1) A shot of one person or one subject; (2) one song, released separately from an album, sometimes used as part of a soundtrack, or a song from an entire score that becomes such a release because of popularity.

SINGLE 8MM FILM

(Also Single 8)

Eight-millimeter film manufactured only in Japan and using a thin base.

SINGLE CONCEPT FILM

A brief, usually silent, film that features a single demonstration, principle, procedure, or idea. (See also Short, Skill Film)

SINGLE/DOUBLE-SYSTEM SOUND CAMERA

A self-blimped camera system that can be used with single- or double-system sound shooting. (See also Blimp, Self-Blimp)

SINGLE FILM SUBTRACTIVE COLOR PROCESS

See Monopack Subtractive Color Process.

SINGLE-FRAME MOTOR

A motor used on an animation camera that runs constantly but has a clutch connected to the drive shaft; when activated by the animation operator, the clutch engages long enough to turn the shaft one revolution or one frame.

SINGLE-FRAME RELEASE

A button, lever, or cable release connection that allows a camera operator to expose frames one at a time.

SINGLE-FRAME SHOOTING
(Also Single Framing)

The filming of one frame at a time used to speed up the motion of objects, persons, or subjects, such as plants, when the film is projected at a normal rate of speed; also used in animation. (See also Fast Motion, Stop Motion, Time Lapse Photography)

SINGLE FRAME TERMINAL

See Frame Stopping Terminal.

SINGLE-PERFORATION FILM
(Also Single Perf Film)

Film with perforations manufactured on only one side; usually one per frame.

SINGLE SCRIM

One layer of scrim netting used for light reduction.

SINGLE SHOT TECHNIQUE

A filming technique of shooting one shot at a time, instead of using the master shot technique. Jump cuts can be avoided by careful attention to the performers' positions at the ends of shots, by using overlapped action, or cutaways. (See also Master Shot Technique)

SINGLE SYSTEM

A simple system that allows for the recording of sound and picture in synchronization, on the same piece of magnetic or optical film, in one camera unit that records both sound and picture; used especially for news footage and amateur filmmaking.

SINGLE-SYSTEM SOUND RECORDING

The recording of sound at the time film is shot and reproducing both in synchronization on the same strip of magnetic or optical film.

SITCOM

Abbreviation for Situation Comedy.

SITUATION

The basic interrelationships and events that are the backbone of a scene, entire production, or film.

SITUATION COMEDY
(Also Sitcom)

An episodic series, pilot or any film or production whose plots have a comic tone and that revolve around continuing characters and their interactions with their particular situations and circumstances. Most sitcoms take place on the same basic set and are taped or filmed in front of live audiences. Some examples of sitcoms are "All in the Family," "The Mary Tyler Moore Show," "Happy Days," and the classic "I Love Lucy."

16MM FILM
A common size film stock, sixteen millimeters wide, available with single or double perforations.

SKED
Abbreviation for Schedule.

SKILL FILM
A brief film depicting how to perform a specific task, shown in deliberate, repetitious steps. (See also Single Concept Film)

SKIN FLICK
A slang expression for films that emphasize nudity; a pornographic film. (See also Pornographic Film)

SKIP FRAME PRINTING
(Also Skip Printing)
An optical printing technique used to reduce the length of a film by having the printer skip every second or third frame.

SKIPPING EFFECT
(Also Strobe Effect, Strobing)
An unwanted jerkiness that appears, especially in perpendicular lines, on screen images and most often occurs in pan shots or shots of objects that move across the screen.

SKULL SHOT
A close-up of a reaction by a performer. (See also Close-up, Shot)

SKY FILTER
A contrast filter, usually yellow, that can be attached to a camera lens to darken only the sky when shooting on black-and-white film. (See also Contrast Filter, Filter)

SKY PAN
A floodlight providing an extremely wide beam of light. (See also Pan, Skylight)

SKYLIGHT (Also Sky Pan)
(1) A bluer, more diffused light reflected from the sky; indirect sunlight; (2) a studio lamp using a bulb of from 5,000 to 10,000 watts, used to light large areas in a studio.

SKYLIGHT FILTER
A pale pink filter used to absorb ultraviolet rays when shooting outdoors with black and white or color film. Not to be confused with a sky filter, which is a contrast filter.

SLAPSTICK
A special sound effect used in early comedy films, where two flat pieces of hinged wood make a loud, sharp clap when struck against an object or performer. This device was used so often in early films that they became known as slapstick comedies. The term is still applied to exaggerated, often silly, comedy. (See also Slapstick Comedy)

SLAPSTICK COMEDY
An almost cartoonlike physical action comedy, violent, acrobatic, and comedic. (See also Slapstick)

SLASH PRINT
See Scratch Print

SLATE
(1) An identification device filmed or taped at the beginning or end of a take on which is written the shot number, take number, and other key production information, used for easy identification of each shot in the lab and in editing; (2) the act of using the slate. (See also Clapboard, Clapstick, Electronic Clapper, Slateboard, "Stick It," Tail Slate)

SLATEBOARD
A small black board used as a slate on which identifying information, such as production number, scene, take number, director and camera operator's name, is chalked. The slate board is photographed at the beginning or end of each take. (See also Clapboard, Clapstick, Electronic Clapper, Slate, "Stick It")

SLAVE
Any device or tool, usually electronic, controlled out of the field of action and out of camera range.

SLED
A bracket mounted on a wall that supports a light.

SLEEPER
A motion picture, usually low budget with little advertising, that opens to little fanfare and on its merit alone becomes a hit and receives unexpected box office or critical success. *Breaking Away* was considered a sleeper and went on to win an Academy Award as Best Picture. (See also Blockbuster)

SLIDE
A photographed image processed on transparent material for projection on a screen. Slide film is manufactured specifically to produce a positive transparency when developed. Slides are often used to project a background in a film or production. (See also Balop)

SLIP FOCUS
See Differential Focus.

SLOP TEST
Processing and developing of a short piece of film outside a lab. Slop testing is done with portable processing equipment as an immediate way to check exposure, registration, and filtration.

SLOW FILM
Film manufactured with a low sensitivity to light. (See also Film Speed)

SLOW LENS
A lens with a maximum f-stop of 2.8.

SLOW MOTION
Action that appears on the screen as slower-than-normal movement, created by filming the action at a faster-than-normal camera speed (overcranking) and then projecting at a standard or slower-than-standard projection speed. (See also

High Speed Camera, Slow Speed, Strobe)

SLOW SPEED
A shooting speed slower than normal, resulting in motion faster than normal during normal projection. (See also Slow Motion).

SLUG
A piece of blank leader inserted into a picture or soundtrack workprint. (See also Leader, Slug In, SMPTE)

SLUG IN
An editing technique that gives desired length to a workprint by inserting blank leader into the picture or soundtrack workprint.

SMPTE
(1) Society of Motion Picture and Television Engineers; (2) SMPTE universal leader—head and tail leader designed and standardized by the Society of Motion Picture and Television Engineers that is appended to release prints to enable projectionists to change over from one projector to another without breaking the continuity of the film being projected. Used in television to thread a film for exact cueing in a program. This leader was introduced in 1965 and has the numbers 8, 7, 6, 5, 4, 3, and 2 each printed over 24 frames, or one appearing each second at standard projection speed. (See also Leader); (3) SMPTE test film—a series of films available from the Society of Motion Picture and Television Engineers used as test footage in equipment; for example, one film allows for the testing of flutter in a projector.

SNAP THE SHOTS
In videotape production, literally the director's fingers snap to signal the change from one camera to the other.

SNEAK
(1) To introduce voices, sound, or music into a film at an extremely low volume; (2) a shortened term for a sneak preview.

SNEAK PREVIEW
A screening of a major release feature film in a commercial theater to get a genuine, broad audience reaction. The audience is sometimes asked to fill out prepared questionnaires giving their honest reactions. Often the picture is reedited after a sneak.

SNOOT FUNNEL (Also Snoot)
A funnel-shaped tube, attached to a spotlight, that controls and limits the width of the beam.

SOAP, SOAPS
Soap Opera.

SOAP OPERA
(Also Soap, Soaper)
Continuing story lines that peak and resolve around the lives of several individuals or families. Usually as one or more stories resolve, others are beginning to peak. In the early days, soap operas were aired daily in half-hour segments and were known for being overly melodramatic in plot and performance, mostly with cliché cliffhangers and second-rate production values; organ music traditionally punctuated scenes. Modern soaps, though still

aired daily in daytime programming hours, are often one hour, with high-quality production values and performances. One of the most popular soaps, created by Agnes Nixon (and often given credit for changing the way soaps are presented), is "All My Children," which features current issues in quality drama, and location shooting (including Switzerland and the Caribbean). Soap operas have been added to primetime programming with such popular once-a-week editions as "Dynasty" and "Dallas." Soap operas were so named because in their early days they were most frequently sponsored by detergent companies.

SODIUM PROCESS
(Also Sodium Vapor
Traveling Matte Process)

A matte made by photographing action in front of a yellow screen, which is illuminated with sodium vapor lamps.

SOF
Abbreviation for Sound on Film.

SOFT
Slightly out of focus, unsharp; images can be seen but appear softened. (See also Soft Focus)

SOFT CUT
Very short dissolve.

SOFT EDGE WIPE
A wipe with a softly blurred edge. (See also Wipe)

SOFT FOCUS
(1) A camera technique used as a style of filmmaking or a special effect in which the subject or background is softened by diffusing the light; using an optical lens attachment, such as a soft focus lens; reducing the focus of a standard lens; or by draping gauze over the lens; (2) an image that is slightly out of focus. (See also Focus)

SOFT LIGHT
A diffused light; light that tends to spread and scatter, such as indirect sunlight or light from a floodlight that produces gradual shading between highlights and shadows. (See also Hard Light)

SOFT PORN FILM
A mild pornographic film in which the content, sexual contact, and nudity is not as explicit as in a standard pornographic picture and in which any sexual acts are merely simulated. (See also Pornographic Film)

SOFTWARE
Commercially available programming and programming materials such as films, videotapes, audio discs, and slides. (See also Hardware)

SOUND
The audio portion of a film or television show comprised of music, voices, or sound effects. Any audible, reproducible tone or recording. (See also Live Sound, Location Sound)

SOUND ADVANCE
The standard difference between a specific point on a soundtrack and the frame of picture that it matches for exact synchronization. The in-

terval is necessary because, at any point, sound and picture are reproduced from different points on a projector. (See also Projection Sync, Sound Displacement)

SOUND ASSISTANT
See Cable Man.

SOUND BOOM
See Boom.

SOUND BOOTH
A small room containing a microphone in which performers record without having their voices interfere with sound or dialogue being recorded on other microphones in nearby sound booths.

SOUND BRIDGE
An audio transition as opposed to a visual. (See also Bridge)

SOUND CAMERA
A camera that is blimped and runs at 24 frames per second. A sound camera achieves synchronization with the sound recorder by either a synch-pulse generator, a synchronous motor, or a crystal. (See also Crystal Motor)

SOUND DISPLACEMENT
The interval or difference in the position of the picture in relation to the soundtracks or release prints. (See also Projection Sync, Sound Advance)

SOUND DISSOLVE
The fading out or overlapping of sound from one track while fading in sound from another. (See also Cross Fade, Dissolve)

SOUND EFFECTS (Also SFX)
All sounds, other than normal dialogue or music, recorded onto a soundtrack. Effects are usually recorded on separate tracks and later, in editing, mixed with the main soundtrack. In addition to such sounds as glass breaking, gun shots, a door slamming, a baby crying, thunder, or crickets, laughter and applause are sound effects often added to the background of situation comedies. (See also Effects)

SOUND EFFECTS LIBRARY
A collection of sound effects recorded and stored in files for easy access on discs, tape, or soundtracks. (See also Library Sound)

SOUND FILM
Film on which sound can be recorded; a film base treated with a magnetic oxide upon which sound is recorded. The sound film is put in sync with the images by matching the perforations on the workprint with those on the sound film.

SOUND HEAD
(1) A magnetic head or light box that picks up sound from magnetic or optical tracks; (2) a magnetic erase head.

SOUND LOG
A printed form, similar in intent to a camera log, upon which the sound person keeps track of roll and take numbers.

SOUND LOOP
An endless loop of audio tape, magnetic film, or soundtrack film containing a sound effect that, when

needed, runs continuously during an editing mix, playing such sounds as crowd noises, birds, traffic, or rain. (See also Loop)

SOUND MASTER POSITIVE

An interim, special sound print made from a sound release negative that will produce sound duplicate negatives from which release prints will be made.

SOUND MIXER

(1) The individual who combines soundtracks into one mix; (2) the controls and console used to combine the soundtracks. (See also Mix)

SOUND ON FILM (Also SOF)

Refers to a camera that can record sound on the film being exposed.

SOUND PERSPECTIVE

The impression of distance in recorded sound; for example, if a train is seen in the distance, its whistle, added in mixing, would have to be at an appropriately lower volume than if the train were a few feet away.

SOUND PRINT

A sound-only positive print.

SOUND READER
(Also Reader)

An optical or magnetic device used primarily during editing to play back soundtracks.

SOUND RECORDER

See Galvanometer Recorder.

SOUND RECORDING

(1) The process of recording sound on tape, film, or disc (See also Variable Hue Sound Recording); (2) the recording itself.

SOUND RELEASE NEGATIVE

The optical sound negative used for the final sound printing on the release print.

SOUND SPEED

The standard speed for filming and projection of sound films; 24 frames per second for all film sizes.

SOUND STAGE

An acoustically treated soundproof building specially built on a studio lot for filming or taping; a soundproof room or production facility that can be used for filming or videotape production.

SOUND STOCK

Film manufactured and used specifically for making films with soundtracks.

SOUND STRIPE

A thin band or stripe of iron oxide in a standardized position on a film strip that allows for the recording of sound with the picture; same as mag stripe or magnetic strips. (See also Magnetic Stripe)

SOUND TAPE
(Also Mag Tape, Magnetic Tape)

Thin plastic tape coated with iron oxide that is used to record sound.

SOUND TRANSFER
The act of rerecording sound from one medium to another, as from disc to tape.

SOUND TRUCK
A truck or van fully equipped for sound recording.

SOUND WORKPRINT
A copy of a soundtrack used in editing and mixing.

SOUNDTRACK
(1) Thin iron oxide stripes located on either or both sides of photographic film on which the audio portion of the film is recorded; (2) any recorded audio tape whether it contains music, dialogue, or special effects, also called track; (3) a copy of the original musical score from a motion picture, usually sold as an album or cassette; a motion picture soundtrack. (See also Buzz Track, Negative Soundtrack, Optical Soundtrack, Positive Soundtrack, Unmodulated Track)

SOUNDTRACK APPLICATOR
(Also Soundtrack Solution Applicator)
Used in the processing of certain film prints as an interim step in which a separate chemical solution is applied to the soundtrack area only.

SOUTH
A directional term used in animation to indicate the lower portion of an animation field chart. (See also Animation)

SPACER
A device placed between reels on rewinders that keeps them aligned in proper feed or take-up positions in a synchronizer.

SPECIAL
(Also Special Programming, Special Program)
Any nonregularly scheduled broadcast, i.e. sporting events such as the World Series or the Superbowl; entertainment events such as the Oscars and Emmys; major news events; or major productions such as "Baryshnikov On Broadway."

SPECIAL EFFECTS
Any visual action, image, or effect that cannot be obtained with the camera shooting in normal operation directly at the action; and which requires prearranged special techniques or apparatus added to the camera, action, processing, or editing. Special effects include contour matting, multiple image montages, split screens, and vignetting; animation; use of models or miniatures; special props such as breakaway glass and furniture; simulated bullet wounds, injuries, explosions, floods, fires, and any mechanical or visual effect whether created on location, in the lab during processing, or in editing in postproduction. (See also Effects, In-the-Camera Effects, Laboratory Effects, Mechanical Effects, Miniature, Out-of-Focus Effect, Painting on Film, Wave Machine)

SPECIAL EVENTS
Entertainment industry occasions, hyped by the press and cov-

SPECTACULAR
(Also Spectacular Film, Spectacle Film, Spectacle)

A feature film whose setting is unusually large and splendid, boasting a cast of thousands and a budget of millions. The subject, generally, is a historical event. *Gone With the Wind, Ben Hur, Cleopatra, Dr. Zhivago*, and more recently *Gandhi*, are classic examples.

SPECTRUM
See Visible Spectrum.

SPECULAR REFLECTION
Regular, normal light reflection.

SPECULATION
See also On Speculation.

SPECULATION FILM
(Also Spec Film)

A film made without a contract, a buyer, or a distributor. The producer of a spec film anticipates recovering his costs and making a profit through rental and sales of the finished product. (See also On Speculation)

SPEED
(1) A command given by the director to start the camera, "roll speed" or "roll'em" indicating the cameras and recorders should begin operation. Sometimes a director calls out "speed" after "roll'em"; (2) a response given by the camera operator after the director has commanded "roll'em" or "speed," to indicate that the camera is running at the correct speed to begin filming. The director will not call "action" until the camera operator tells him that there is speed; (3) exposure index or light sensitivity of film; (4) the lowest f-stop or t-stop of a lens that expresses the largest amount of light a lens is capable of transmitting.

SPEED LINES
Painted lines appearing behind animated figures indicating fast forward motion.

SPHERICAL ABERRATION
A defect in a lens that causes square images to appear circular in the reproduced image.

SPHERICAL OPTICS
A lens with a spherical surface as opposed to an anamorphic lens.

SPIDER
See Tee.

SPIDER BOX
(Also Plugging Box, Tie-in Box, Octopus Box)

A conveniently sized, portable, electric terminal with outlets into which lights or other equipment may be plugged.

SPIDER DOLLY
A wheeled dolly whose wheels are on projecting legs.

SPILL LIGHT
An unwanted lighting effect caused by the scattering of light along the main beam of a light source.

SPIN
A special effect produced on an optical printer or animation stand in which the image whirls or spins.

SPINDLE
A film rewinder's shaft; a shaft that projects from a film rewinder.

SPIRAL WIPE
See Wipe.

SPLASH LIGHT
Unwanted light on a rear projection screen that comes from the light source used to illuminate the performers.

SPLICE
(1) The permanent joining of two cut pieces of film or tape by one of several techniques (See also Splicing); (2) the point at which two pieces of film or tape are joined by splicing.

SPLICER
A machine for joining together pieces of film, videotape, or audio tape. There are cement splicers, tape splicers, hot or cold splicers, negative or positive, hand- or foot-operated splicers in varying degrees of sophistication from tiny, portable, hand-held units to more complex devices used in editing rooms. The main function of a splicer is to hold the film or tape still and in proper alignment when cutting or joining it to another piece of film, film leader, videotape, or audio tape.

SPLICING
The process of making a splice; the following techniques are most common: cementing, butt-welding, taping, staples, gommets (used only in processing), or overlapping. Splicing is usually done by using a splicer but can be accomplished without one by simply cutting the film or tape with a razor or scissors and applying splicing tape or cement to join the cut end to the desired leader or strip of film or tape. (See also Butt Splice, Butt-weld Splice, Cement, Overlapping, Positive Splice, Straight Cut, Tape Splice)

SPLICING BLOCK
A small device that firmly holds the ends of a film or tape to be cut or joined. Most blocks have a vertical groove to guide a razor blade in cutting picture film and a diagonal groove for cutting magnetic tape. The block also holds the film or tape in alignment while cement or tape is applied.

SPLIT FIELD LENS
A lens capable of focusing on two different points. Because the lens is half close-up and half plain glass, it is able to focus at the same time on a point close to the camera, and on one further away. (See also Close-up Lens, Lens)

SPLIT FOCUS
The technique of bringing two subjects into sharp focus when one is close to the camera and the other is further away. (See also Differential Focus, Focus, Split Field Lens)

SPLIT FOCUS SHOT
A shot in which the focus is changed from one point near the

camera to another further from the camera, or vice versa.

SPLIT SCREEN

(1) A technique used to duplicate an actor or an object in a scene, for example, when one performer plays two roles. This technique is achieved by dividing the film frame into separate images in the camera or in an optical printer; (2) an effect created by any method on videotape or film in which the screen appears divided two or more times, with two or more separate images displayed. For example, if a rock group is playing, the drummer can appear in close-up on one corner of the screen; the keyboard player in another, guitarists in other corners, while a lead singer is vignetted in the center of the screen. In split screen technique, each image is a separate shot or repetition of a separate shot. The shots appear on the screen at the same time.

SPONSOR

(1) A person, group, or organization for which a program or film is made by contract; or one who buys all or a portion of it after production is complete; (2) a corporation, organization, or the manufacturer of a product who pays a network or distributor for advertising before, after, or during a specific production, which can offset the cost of the production as well as generate revenue for the network or distributor. Hence, the often heard phrase on television, "brought to you by ..."

SPOOL

A device around which film is rolled. If it is used for daylight loading, this spool will be equipped with close-fitting, light-tight flanges that protect the film from light. A spool with the flanges cut into pie-wedge portions is usually referred to as a reel.

SPOOL BANK
(Also Roller Bank)

Groupings of rollers over which a film loop travels.

SPOOL CAPACITY

Refers to the size of a camera's film chamber and the largest daylight loading spool it can accommodate.

SPORTS

A category in television, usually commercialized, such as the showing of games or athletic specials (the Super Bowl or the Olympics) for commercial broadcast. Also, a segment on most television news shows.

SPOT

(1) A bit part, also a portion of a production, for example, the music or comedy spot of a variety show; (2) a commercial announcement; (3) commonly used colloquial term for a spotlight; also the concentrated circle of light formed by a spotlight; (4) "Spot"—the command to have the spotlight follow a performer or subject.

SPOT BRIGHTNESS METER
(Also Spot Meter,
Spot Exposure Meter)

A light meter that measures reflected light at an extremely narrow

SPOT NEWS
A scheduled television mini-cast, usually five minutes or less in duration.

SPOTLAMP
A bulb for a spotlight; a general term for any of numerous studio lamps of similar design but different in size, wattage and properties, for example, halogen lamp. (See also Lamp, Halogen Lamp, Quartz Lamp, Spotlight, Tenner)

SPOTLIGHT (Also Spot)
(1) A light that projects an adjustable, focusable, and relatively narrow beam of illumination. By using a Fresnel lens and reflector the spot can be adjusted from a concentrated circle of light to a flood beam pattern; the most commonly referred to sizes are baby, junior, and senior. (See also Baby, Junior, and Senior); (2) the command to "spot" or "spotlight" a subject; (3) a lamp that produces a spotlight (see Spotlamp). (See also Tenner)

SPOTTING
Undesired marks or spots appearing on film due to faulty processing.

SPOTTING IN
The proper placement of music and/or sound effects in the desired position of a tape or film.

SPRAY PROCESSING
A film processing technique in which the film is sprayed with chemicals, as opposed to using a bath.

SPREAD
A term of measurement referring to the width of a beam of light.

SPREADER
See Tee.

SPRING DRIVE CAMERA
A camera powered by a spring drive motor.

SPROCKET HOLES
See Perforations.

SPROCKET ROLLER
(Also Sprocket Wheel)
A device consisting of a roller with teeth, over which the perforations of a film fit, to pull the film along in a camera, projector, or printer.

SPROCKET TOOTH
(Also Sprocket)
One of the teeth, or projections, of a sprocket roller.

SPY FILM
A film genre whose central theme is espionage.

SQUAWK
A slang expression for a limited or low quality speaker system.

SQUEEGEE
A soft sponge or wiping blade used to remove moisture from film.

SQUEEZE
The act of compressing film images horizontally by using an ana-

morphic lens. (See also Cinemascope)

SQUEEZE RATIO
A term of measurement referring to the amount of horizontal anamorphic compression of an image in relation to its height.

SS
Abbreviation for Stock Shot.

STAGE BUSINESS
See Business.

STAGE MANAGER
Stage manager, a position governed by the DGA, is the sole liaison between the director or assistant director and all persons or facilities in the studio, stage, or shooting location. Specifically, here are some of the duties performed by a stage manager: Supervises all stagehands, chief carpenters, and prop people in marking and/or cueing performers, their positions, scenery placement, props, special effects, drops, curtains, travelers, or similar devices. A stage manager gives all cues from the floor and relays information, such as stage directions from the director or assistant director, to all persons on the stage, or location. The stage manager also designates dressing rooms, notifies performers on the set of rehearsal times, and gives break announcements.

STAGING
Creating the blend of blocking, sets, costumes, lighting, and action for a production. (See also Mise-En-Scène)

STAND
An adjustable pedestal used to support a light, gobo, scrim, or reflector; also made in a style to display music, pages of script, or production cues.

STAND-IN
A person who physically resembles an actor or actress and who replaces that performer while lights and equipment are being adjusted and tested. In addition, certain stand-ins replace performers in scenes that are brief and distance shots, usually on location, as well as sometimes in difficult or uncomfortable shots in which perhaps only the performer's back is visible. For example, if a bucket of water is thrown at a performer and several takes are needed, the performer would possibly be used only in a reaction shot. Stand-ins are not used for extremely difficult or dangerous shots; stunt people are usually called in for that purpose. (See also Double)

STANDARD CANDLE
See Candle.

STANDARD GRAY CARD
See Gray Card.

STANDARDS
(Also American Standards)
Specifications printed on film and related equipment established by the American National Standards Institute to standardize communication of performance capabilities, etc.

STANDING SET
A set that has been assembled and is ready for production; also a constructed set that stands upright, as opposed to a natural outdoor location setting.

STANISLAVSKI METHOD
See Method Acting.

STAR
The commonly used word for a well-known performer; a celebrity.

STAR FILTER
A filter that produces an image of radiant beams appearing to extend from highlights in the field of action. (See also Filter)

STAR SYSTEM
(1) An outdated method of creating a star through media hype and extensive publicity stunts; (2) a colloquial expression for the fact that stars can command extremely high salaries and production extras (such as limousines and deluxe dressing rooms) because of their box office draw. (See also Box Office Draw)

STARRING
A billing credit in screen titles to indicate a leading or key role in a production. (See also Billing, Credits, Titles)

START MARK
A film frame that is specially marked to indicate the beginning position for any procedure where an exact commencement is required, such as printing, rerecording or interlock projection.

STATIC
Audio interference that is heard as crackling or popping noises in any sound system or broadcast signal.

STATIC BACKGROUND
(1) Same as still background (See also Still Background); (2) a term sometimes used to describe audible static in the background of a soundtrack. (See also Static)

STATIC MARKS
(Also Static Electricity Marks)
Unwanted lines or marks on film, caused by rewinding film too quickly or at an uneven speed (especially in low temperatures and low humidity conditions), or by friction as the film moves through equipment, discharging a spark as the edges hit certain points in cold, dry conditions.

STEADICAM
A trade name for a commonly used camera support system that attaches to the operator's body and allows for free movement by the operator without causing the picture to jump. A device holds the camera steady while the operator moves, and uses a small television monitor as a viewfinder. A Steadicam frequently is used on location, and in action shots to produce a feeling of reality.

STEADINESS TEST
A test that checks for defective internal camera movement that can result in unwanted wobbling or motion on the processed film. A steadiness test is not routinely given but is

performed, for example, when a camera has had major repairs.

STEP-CONTACT PRINTER

A contact printer in which the raw stock and original film are advanced intermittently frame-by-frame. The raw stock is exposed to the printer light only when both films are stationary. (See also Contact Printing)

STEP LENS

A condenser lens with concentric prisms on the plano or flat side. (See also Lens)

STEP OPTICAL PRINTER

An optical printer (as opposed to a contact printer) that exposes the film one frame at a time. (See also Optical Printer)

STEP PRINTER

Any printer, contact or optical, that holds each frame still during exposure.

STEP PRINTING

A technique of printing a film frame-by-frame in which the printer is stopped between frames. Step printing can be used to print a film whose original is in poor condition and that would be impossible to duplicate on a continuous printer.

STEP-PRISM
(Also Stepped Prism)

A lens such as a Fresnel lens that acts as a glass refraction device and whose surface is composed of prism wedges varying slightly from the center to help form a tight beam of light as the light comes from a lamp filament.

STEP WEDGE
(Also Step Tablet)

A test strip of developed exposures made by a sensitometer in order to evaluate the development process. (See also Test-run)

STEREO

(1) A three-dimensional quality of reproduced sound created by two or more separate sources or tracks of sound; (2) a system for recording or playing back in stereo; (3) a positive transparency that is used for the projection of a rear-screen background.

STEREOPHONIC SOUND

The combining of sound and perspective in an audio recording or playback unit with the use of two or more microphones (each with its own track) and the use of multiple tracks that will be mixed down into two basic areas of sound, which achieves a three-dimensional or live effect in the sound when played back.

STEREOPHONIC VARIABLE-AREA TRACK
(Also Stereo Variable-Area Track)

A variable-area sound track with a double pick-up area on which each area of the track is capable of recording and reproducing sound from an independent microphone.

STEREOSCOPIC CINEMATOGRAPHY
(Also 3-D, Stereographic Cinematography)

Motion picture photography that produces images that appear three-dimensional to the eye of the viewer. Two images are produced, each seen by a different eye of the viewer (the left image is seen by the left eye and the right image by the right eye) and given depth and dimension when viewed through special glasses. (See also Three-Dimensional, Stereoscopic Film)

STEREOSCOPIC FILM

An experimental method to achieve the appearance of depth. The film is exposed in matching pairs of frames with one frame shot of the left-eye view of a scene, and the other frame shot of the right-eye view of the same scene and then projected together, side by side. There is a slight but definite difference in the angle of view, or parallax, between the two sets of frames, which provides a realistic depth perception when the left-view frame is presented to the left eye, and the right-view to the right eye.

STET

(1) "Stet"—dialogue, sets, props, or anything in production that remains when other things are changed or edited out; (2) a show's opening or closing that has the same format every time it is aired, i.e., stet opening or closing.

"STICK IT"
(Also "Sticks It," "Slate It")

The command to hold the slate in front of the camera and clap the sticks together, a task usually performed by the stage manager. Shot and take numbers may be called out at the same time. (See also Clapboard, Clapstick, Electronic Clapper, Slate, Slateboard)

STILL BACKGROUND
(Also Static Background)

A background of a scene with no motion or action of any kind.

STILL MAN

A photographer assigned to a film or television production crew to take still photographs during the production. (See also Still)

STILL/STILLS

A nonmoving picture, a standard black-and-white or color photograph; stills are often used to publicize an actor or film and are published in newspapers and magazines, as well as being displayed in theater lobbies before a picture opens and during its run. (See also Still Man)

STING

(1) A stinger; (2) the instruction to add, or the process of adding, a stinger to a soundtrack. (See also Stinger)

STINGER

Anything used to punctuate or accentuate a musical or special effects soundtrack, for example, a sudden burst of sound, musical phrase, or chord. A stinger can be a final drum beat or series of final dramatic musical chords; but it is usually brief and adds a definite end

STOCK

to a soundtrack whether created electronically, musically, or from any sound source.

STOCK
(1) Unexposed, raw stock; (2) a shortened term for stock footage (See also Stock Footage); (3) a traveling or short-term production, such as summer stock.

STOCK CHARACTER
A clichéd, stereotypical character such as the standard villain in black with a moustache and beady eyes, or the sweet little old grandmother with a bun and glasses in a rocking chair. (See also Character)

STOCK FOOTAGE
(Also Stock Shots, SS)

Prefilmed or taped library footage of sequences, incidents, scenery, and action that can be inserted generically into any production as desired, for example, a plane taking off, an ocean sunset, an historical event, or footage of famous places. Some shows have their own stock footage that is reused in almost every episode. Examples are "Love Boat," which has stock footage of various shots of the boat at sea; and in *Battlestar Galactica*, in which certain explosions are repeated from stock footage. (See also Library Shot, Stock Footage Library, Stock Shot)

STOCK FOOTAGE LIBRARY
The place where stock footage is collected or stored, with a filing system for easy access.

STOCK PART
A part in which one plays a stock character.

STOCK SHOT (SS)
A scene or clip taken from a stock footage library, as opposed to a shot made for that particular production, usually of hard-to-duplicate action or scenic locales. (See also Stock Footage, Stock Footage Library)

STOP
The lens opening or aperture measurement in terms of the f- or t-scale. (See also F-Stop, T-Stop)

STOP BATH
A chemical bath that halts the action of a film developer, usually containing acetic acid.

STOP DOWN
To limit the amount of light entering the lens by reducing the size of the aperture, for example, changing the f-stop from 5.6 to 8. Also, to use a stop lower than normal for a shooting situation.

STOP MOTION
(1) When referring to animation, the technique of exposing each individual frame separately; (2) a special visual effect accomplished by having the performers freeze in place during a shot, while something is changed on the set, such as adding or removing a prop or person, resulting in the illusion that an object or a person has suddenly appeared or disappeared; (3) sometimes used to refer to time-lapse,

fast motion, and single-frame shooting.

STOPWATCH
A standard instrument used on most productions to time the action, usually calibrated in feet and frames for motion picture work. In television production, the p.a. (production assistant) keeps time. In film, the script person usually does. (See also Timing)

STORY
The basis of a script, usually in the form of a written narrative, but it can spring from verbal exchanges in creative story conferences.

STORY BOARD
A step in the development of a script whereby a series of drawings are laid out, picturing the key scenes or shots. Often, beneath the sketches, are indications of dialogue, music, special effects, or camera angles.

STORY CONFERENCE
A meeting of the writers and producers to discuss a script, its direction, characters, and possible revisions, or, to create new story ideas or directions for an episodic television show or for a new project.

STORY EDITOR
In television, one who reads written scripts to be certain that all dialogue and actions are consistent with each character. (See also Bible) In film, one who reads stories being considered for production and sometimes edits stories before they are developed into shooting scripts.

STORY LINE
The main connecting thread, sequence, or plot of a script. (See also Outline, Script, Treatment)

STRAIGHT
A line or lines delivered without revealing give-away emotion or inflection, such as a set-up line for a joke.

STRAIGHT CUT
Refers to the technique of butting two shots together, without using an optical effect. (See also Butt Splice)

STRAIGHT MAKEUP
Natural looking makeup that only serves to counteract the effect of stage lighting; the performer's features are enhanced, but not altered as in the application of character makeup. (See also Makeup)

STRAIGHT MAN, STRAIGHT PERSON
A person whose key role it is to set-up lines for another performer's jokes, punch lines, or humorous conversation. In comedy teams, one is usually the straight person for the other; for example, Dick Smothers is often the straight man for brother Tom's humorous situations and punch lines; Johnny Carson is a master straight man for his guests; however, Ed McMahon often takes on the role to set up some of Johnny's jokes.

STRETCH OUT
To lengthen dialogue or narrative by adding pauses and delivering at a slower pace in order to fill time, as ordered by the director.

STRIKE
(1) To tear down a set and store the props, flats, etc., upon completion of production, "strike the set" is the commonly used phrase; (2) to make a duplicate of a film or tape, "strike a copy."

STRINGOUT
See Assemble.

STRIP TITLE
Superimposed titles or images that travel horizontally across the screen.

STRIP/STRIPPED SHOW
A syndicated television show that airs five days a week.

STRIPE
A magnetic stripe on a film track.
(See also Magnetic Stripe)

STRIPLIGHT
(Also Border Light)
A type of floodlight, consisting of a row of lights contained in a narrow three-sided box, generally used for cycloramas. (See also Floodlight)

STROBE
A device that allows a light to skip or flash so quickly that, to the eye, the subject illuminated appears to be moving in slow motion. (See also Slow Motion) Also, a flash attachment commonly used on still cameras.

STROBE EFFECT
A lighting effect created by a strobe. (See also Skipping Effect, Strobe)

STUDIO
(Also Film Studio, Tape Studio)
(1) A usually large, acoustically prepared, enclosed area for tape, film, or live production. Live and tape studios frequently are smaller than those used for film production. Live news broadcasts and episodic television shows require a smaller studio, than does filming of a major motion picture, such as *The Best Little Whorehouse in Texas*, which shot in the largest studio on the Universal lot; (2) a large production company, usually one with its own studio lot (see also Lot, Studio Lot); (3) a production facility.

STUDIO CAMERA
Any camera manufactured to operate in controlled studio conditions and used strictly for studio work because of its large size and weight.

STUDIO EXTERIORS
Settings that on screen appear to be outdoors, but that are actually shot inside a studio; also refers to the shots filmed or taped at such locations. (See also Exteriors)

STUDIO FILM
A film made entirely in a studio or soundstage.

STUDIO LOT
The general term referring to the studio location itself, that houses production facilities and often the

home office of a distribution company. Usually refers to a major studio, which includes sound stages, executive offices, production offices, back lot and areas designated for sets, props, makeup, and wardrobe. The major Hollywood studios that surround the Los Angeles area form the largest and most well-known group of studio lots and include: Twentieth Century-Fox in Century City; Paramount in Hollywood; Warner Brothers and Columbia in a shared facility at the Burbank Studios; Buena Vista (Disney Studio) in Burbank; MGM in Culver City, which also houses United Artists West Coast production headquarters, and Universal Studios, which is so large that it is designated as its own city, Universal City, California. (See also Lot, Studio)

STUNT (Also Gag)

A complicated and often dangerous feat, carefully planned and choreographed by a stunt coordinator, and executed by a stunt person. Generally performers (whether stars or supporting players) do not do their own stunts, but are doubled by professional stunt people. Some well-known exceptions to this industry rule are James Garner, Steve McQueen, and Errol Flynn who defied studio heads and performed their own stunts.

STUNT COORDINATOR

The crew member who creates, styles, choreographs, and is in charge of all stunts needed in a motion picture.

STUNT MAN, STUNT WOMAN

The specially trained man or woman who performs dangerous feats in place of performers.

SUBBING LAYER

An adhesive layer binding a film's emulsion to its base.

SUBJECTIVE CAMERA

A filming technique whereby audience involvement is heightened by showing the protagonist's point of view. The camera takes the place of the actor. If a character is being chased through the woods by a wild animal, rather than simply recording the subject running, it is more exciting if the camera operator runs with the camera seeing what the character would see as he goes tearing through the woods, occasionally looking back, and being slapped in the face by branches.

SUBJECTIVE CAMERA ANGLE

A camera angle that allows the audience to see what the subject or performer is seeing. (See also Subjective Camera)

SUBTITLE

A title or lines of dialogue that appear at the bottom of the screen, used generally to translate dialogue in foreign films or to identify the date or locale of a scene.

SUBTRACTIVE PROCESS

Most monopack color films in use today use precisely controlled elaborations of the subtractive process, which is a method of analyzing and resynthesizing the colors on film by using three light filters, each representing a primary color. First a black-and-white negative is made for each primary color, and for each of these "color-separation" negatives, a dye in a color complementary to each filter is used. Superimposition of the three dye images yields a combined transparent positive image in natural color. (See also Additive Process)

SUCCES D'ESTIME
A film or production that receives praise from the critics, but is not well received by the public.

SUCTION MOUNT
See Limpet Mount.

SUN GUN
A colloquial term for a booster lamp added to portable camera equipment to augment existing lighting conditions, especially used by mobile news crews.

SUNLIGHT
Natural light from the sun; daylight. (See also Daylight)

SUNSHADE
A lens hood that protects a lens from glare or the elements.

SUPER
A commonly used shortened term for a superimposure or the instruction to superimpose.

SUPER 16
Sixteen-millimeter film whose extended frame covers most of the soundtrack area, which allows for a wider screen aspect ratio without using horizontal masking.

SUPER 8MM FILM
(Also Super 8)
Film that has 72 frames per foot, with one perforation opposite the middle of the frame. Super 8 film frames are considerably larger than standard eight-millimeter film frames.

SUPERIMPOSE
(Also Double Exposure, Super)
Exposing two or more images on the same frame on a strip of film or tape by recording or filming one image and then rerecording the new image. The effect can be achieved by double exposure (in a camera) or during printing (double printing).

SUPERIMPOSED TITLES
Titles or credits that appear over another picture, whether still, film, or tape.

SUPERIMPOSURE
The result of exposing two or more images on the same frame on a strip of film or tape. (See also Superimpose)

SUPPORT
See Base.

SUSPENSION OF DISBELIEF
The audience's belief in and temporary acceptance of obviously contrived actions or events when presented in the context of a plot. In the opening of *Heaven Can Wait* the audience must accept that Warren Beatty's character is in heaven, while a new body is being found on

earth for him, because he died before his time.

SWEETENING

A soundtrack mixing session in which final sound mixes are completed, especially, in television comedy shows, the augmentation or addition of laughter and applause to the soundtracks.

SWISH PAN
(Also Blur Pan, Flash Pan, Flick Pan, Whip Pan, Zip Pan)

A quick pan or camera movement so rapid that the images are blurred. Sometimes used as a transition between scenes. (See also Pan, Shot)

SWITCH

(1) A videotape term, to cut from one camera to another, used especially in television taping of shows produced before a live audience; (2) "Switch"—the command to switch the on-air video image from the video in one camera to that of another camera, which is positioned at a different angle.

SWITCHBACK

The return to previous action from a cutaway.

SWITCHER

The control room engineer who switches from one picture (camera) to another, as the director orders. (See also Switch)

SYNC

The commonly used term meaning "in synchronization."

SYNC BEEP (Also Sync Tone)

Used in double system shooting with certain cameras, the simultaneous recording of a tone (by magnetic tape recorder) at the exact time when the camera exposes a few frames of film. The light produces a "fogged" section of film that is later aligned with the beeptone, resulting in synchronization of sound and picture.

SYNC MARK

A mark on the film leader that acts as a reference point for synchronization of two or more films; also a mark placed on the frame where the clapsticks are sounded on the slate.

SYNC MOTOR
(Also Synchronous Motor)

A camera or projector motor that can be run electrically or manually at a constant speed and at the same speed as the sound-recording and reproducing machines to which it is synced.

SYNC POP

See Blip Tone.

SYNC-PULSE

A timing device that is keyed to the camera motor speed. This synchronizing signal, or sync-pulse, is recorded with the audio track on magnetic tape and is used as a time reference to maintain synchronization when the audio is transferred.

SYNC-PULSE CABLE

A cable that transmits sync-pulse signals from a camera to a tape recorder.

SYNC-PULSE GENERATOR
A small generator within a camera that produces a sync-pulse signal that is in turn fed into a tape recorder.

SYNC-PULSE SYSTEM
Any sound recording system that records a synchronizing signal on magnetic tape as a timing reference. (See also Sync-Pulse)

SYNC PUNCH
A hole punched in the film leader acting as a reference point for synchronization with other strips of film.

SYNC-TONE OSCILLATOR
An electronic device in a camera that produces a sync beep on the sound recorder to indicate precise speed of the camera motor. These tones are inaudible on normal playback and do not interfere with sound recording; however, in editing they are a guide to exact synchronization of sound and picture.

SYNCING DAILIES
Assembling the picture and sound workprints of that day's shots for synchronous interlock projection. (See also Dailies, Synchronize)

SYNCHRONISM
The quality of being in synchronization or synchronous. The relationship and coordination of picture and sound. There are many devices created to maintain synchronism in a picture so that all sounds and images are precisely connected.

SYNCHRONIZATION
A precisely simultaneous action or image with its sound; an exact match between picture and sound; spoken dialogue that matches the lip movement on the picture. (See also Lip Sync, Out of Sync)

SYNCHRONIZE
The act or process of matching the sound exactly to the action, such as dialogue that precisely matches lip movement or a door slamming and the sound occurring at the same moment. (See also Synchronization)

SYNCHRONIZER
A system with a rotary shaft and sprocket wheels, which holds identical lengths of picture and sound films simultaneously, allowing for two (or more) films to be run in synchronization during editing.

SYNCHRONIZING SIGNAL
An electronic tone or pulse recorded as a control signal on the soundtrack by a generator in the camera.

SYNCHRONOUS
Descriptive of sound and picture that occur at the same time in synchronism, such as footsteps in dry leaves with the synchronous sound of the leaves crackling.

SYNCHRONOUS MOTOR
See Sync Motor.

SYNCHRONOUS SOUND
(Also Sync Sound)
(1) All sounds in the finished picture that are in proper relation to

the action, in the sense that there is an image corresponding to each sound; (2) sound recorded in synchronization with the picture, either in original filming or in post production lip-synching dialogue. (See also Lip Sync, Post-synchronized Sound)

SYNCHRONOUS SPEED

The standard camera speed of 24 frames per second synchronized with the soundtrack.

SYNOPSIS

A complete step-by-step summary of the events or plot of a script, film, or teleplay. A brief synopsis would be a much-condensed version, such as is often printed next to the title of a show's name in *TV Guide*.

SYNDICATION

A show that does not air on one of the networks, but is sold to independent stations around the country.

7

T & A
An abbreviation for the term Tits and Ass, which refers to a type of television program or motion picture whose story deals with titillating subject matter. (See also Factor, Jiggle)

T CORE
A plastic core, 2 inches in diameter, used to hold up to 400 feet of 16mm film.

T-STOP
A system of lens calibration that is considered the true or effective f-stop. F-stops are based on the amount of light entering a lens; T-stops are based on the intensity of the light that emerges from the back of the lens and forms the image. The T-number represents the f-stop number of an open circular hole or of a perfect lens having 100 percent axial transmission. T-stops are calibrated by measuring the light intensity electronically at the focal plane, whereas f-stops are calculated geometrically by dividing the focal length of the lens by the diameter of the diaphragm. There is no fixed ratio between T-stops and f-stops that applies to all lenses. T-stop calibration is especially important with zoom lenses whose highly complex optical design necessitates exact optical elements.

TABLE READING
See Read Through.

TACHOMETER
A meter that registers camera speed in frames per second.

TAG (Also Tag Line)
(1) Information added or "tagged" onto the end of a prepared spot announcement. The information can be audio only, video only, or both. The additional information is usually local times, dates, and places added to the end of a national commercial. For example, in an advertisement for a film, the name of the local theater where it is playing would be tagged onto the spot; (2) a short segment added to the end of a television program that wraps up any loose ends of the plot, not clearly handled at the climax. (See also Exit Line, Outro)

TAIL
See Foot.

TAIL LEADER
Leader spliced onto the end of a piece of film which protects the image on the film as it passes through the projector.

TAIL-OUT (Also Tailsout)
Film wound in reverse with the end of the film resting on the outside or beginning of the roll.

TAIL SLATE
(Also Upside-down Slate)
A slate marked at the end of a shot

instead of at the beginning, when it might be distracting. The slate is usually turned upside down to indicate that it is a tail slate. A tail slate is often used in documentary shooting or in films where the actors are amateurs or children. (See also Slate)

TAILGATE
An optical printer's projector.

TAKE (Also Shot)
(1) One complete scene with no breaks in the action; (2) each repetition of a shot until the director is satisfied with the results. The takes are sequentially numbered on the slate and by voice on the sound track for later identification in editing. For example, take 17 would be the seventeenth time the scene was shot; (3) a quick reaction; a look or facial expression. A double or triple take would be the same reaction briefly reexpressed two or three times.

"TAKE IT" (Also "Print It," "Save It")
A director's command given at the end of a shot indicating that the shot is to be processed into the workprint.

TAKE UP (Also Take-up Reel)
Any device, mechanism, or reel onto which the film is wound after it has passed through the viewing apparatus of a camera, projector, or editing unit.

TAKE-UP SPOOL
The unit in a camera on which the exposed film is wound. (See also Take Up)

TAKING LENS
The lens used to film a scene from a specially designed turret holding two or more lenses. (See also Lens)

TALENT
Any performer paid to appear in a production, including actors, singers, or dancers.

TALENT AGENCY
See Film Talent Agency

TALKING HEADS
A colloquial expression meaning a medium-close shot of people who are talking, with no other action in the shot. (See also Camera Talk, Shot)

TANK SHOT
Action filmed in a large water tank, often involving miniatures of boats, or submarines. (See also Shot)

TAPE
(1) Any magnetic recording tape used for audio and video production. The tape records the sound and/or picture instantly and can be immediately replayed without needing a developing process (See also Videotape); (2) the act of shooting or recording on tape; to "tape" a scene, which means magnetic tape is being used, as opposed to film; (3) loosely, a demo audio or video production.

TAPE RECORDER
A machine that records sound and/or picture on reels or cassettes of magnetic tape.

TAPE SPLICE
A butt splice joined by a short piece of splicing tape. (See also Splice)

TAPE SPLICER
A device that cuts film or magnetic tape to join the ends to other pieces of film or tape, using splicing tape.

TAPE STUDIO
See Studio.

TAPPING THE TRACK
Marking the musical beats on a moving sound track that assists synchronization of other added sound effects in editing.

TARGET
A black disk, 10 inches across or smaller, that, when attached to a luminaire, controls the amount and configuration of the light beam, thus producing the desired shadows. (See also Dot, Finger, Flag, Gobo, Scrim)

TARGET AUDIENCE
The group for which a motion picture or television show is intended, or for whom it would have particular appeal.

TECHNICAL
(1) The mechanical aspects of production involving the technical crew, including camera crew, sound, and sound effects; (2) an acting technique in which the actor plays a scene relying on mechanical technique, as opposed to emotional involvement, without letting the audience see this method. Usually this implies only an adequate performance.

TECHNICAL ADVISER
A person considered an expert in a particular field, hired by the production company to consult on a film or television show. For example, a representative of the Los Angeles Fire Department acted as technical adviser on "Emergency."

TEE (Also Tie-Down, Triangle, Spider, Spreader)
A device made of wood or metal, shaped like a T or Y, used to stabilize the legs of a tripod to keep it from slipping.

TELECINE
See Film Chain.

TELEPHOTO DISTORTION (Also Telephoto Effect)
The effect created in shots made with a telephoto lens, whereby perspective is distorted. Objects far away from the camera appear closer, and action moving either toward or away from the camera seems slowed down.

TELEPHOTO LENS
See Long Lens.

TELEPLAY
A script written for television; a play, book, or film adapted for television.

TELEPROMPTER
Trade name of a commonly used device that replaces cue cards. Performer's dialogue is easily visible on this apparatus, which readily attaches to, or is placed near, a camera.

TELETHON
A live television broadcast, many

hours in duration (sometimes twenty-four consecutive hours), usually for the purpose of raising funds for a charity.

TELEVISE
To broadcast, either live or prerecorded, on television.

TELEVISION
(1) The transmission of audio and video simultaneously as electrical signals or radio waves; (2) the device that receives audio/video signals or waves and translates them onto a viewing screen; (3) the industry that creates and broadcasts programs in this medium.

TELEVISION COMMERCIAL
An advertisement, a short piece of film or tape aired on television to promote a sponsor's product or service, and for which the sponsor has paid.

TELEVISION CUTOFF
(Also TV Cropping)
Cropping of the film frame that occurs during television transmission.

TELEVISION MASK
(Also TV Mask)
A mask used on a viewfinder indicating the television safe-action area; that is, the area safe from cropping during television transmission. Especially used with titles so that all letters are certain to appear on the screen.

TELEVISION SAFE-ACTION AREA (Also Safe-action Area)
(1) The film's frame area that will be intact, that is, not cropped during television transmission; (2) that safe area of a motion picture camera's viewfinder marked by reticle lines or masked to define the area that will definitely appear on the film.

TEMPLATE
An opaque sheet with cutout patterns that, when inserted inside a spotlight, will create the desired shadow effect.

TENNER (Also 10K)
A studio spotlight with a 10,000-watt bulb. (See also Spot)

TEST
The filming or taping of a scene using the actor(s) being considered, as an aid in final casting. (See also Audition, Camera Test, Sides)

TEST FILM
A small quantity of film specifically used with a given piece of equipment to ascertain whether the equipment is operating satisfactorily.

TEST-RUN
A trial operation of a film processing machine to develop a step wedge. This determines if the machine is operating satisfactorily. (See also Stop Wedge)

TEST STRIP
A short piece of film, taken from either the beginning or end of a roll, used to determine whether conditions for filming or processing are as desired.

THEATRICAL FILM
A feature film made specifically

for release in commercial theaters, as distinguished from a movie made for television.

THEME
The main story from which the script is developed and written.

THEME MUSIC
A musical score recurring throughout a film.

THEME SONG
A song written and performed for a specific show, or as a motif for a show, replayed each time the show airs.

35MM BLOW UP
The enlargement of small gauge film, such as 16mm or 8mm, to 35mm film.

35MM FILM
A specific type of motion picture film, 35mm wide, each frame of which has four perforations on each side of the film.

THREAD
(Also Threading, Lacing)
The act of placing the film in a camera, projector, or printer, correctly weaving the film around the spools and rollers of the threading path, in order to use the film.

THREADING PATH
The proper course over which a strip of film will travel through a piece of equipment, such as a camera or projector.

THREE-COLOR PROCESS
The use of all three primary colors in the processing of any film. (See also Color Separation)

THREE-DIMENSIONAL FILM (Also 3-D, Stereoscopic)
Film that seems to have depth and dimension. Characters and props appear to "reach out" toward the audience. This effect is achieved with the use of stereoscopic film viewed through polarized lenses. (See also Stereoscopic Cinematography, Steorescopic Film)

3-EMULSION FILM
See Tripack

360° PAN
A shot (pan) during which the camera revolves 360° on its axis, thus making a complete circle and returning to the point from which it began.

THREE-QUARTER ANGLE
Positioning the camera in such a way that approximately 75 percent of the subject is facing toward the camera, as opposed to a full or frontal shot. (See also Camera Angle, Shot)

THREE SHOT
A shot framing three performers. (See also Group Shot)

THREE-TRACK STEREO
Sound recorded with a three-microphone pick-up pattern or through multi-microphone and line-type circuits in which the reproduced sound has a distinct left, center, and right channel of distribution.

THREE-WALL SET
A set comprised of three walls

that when assembled forms two corners. (See also Set)

THROUGH-THE-LENS REFLEX (Also TTL)
Refers to a camera that allows focusing through its lens without parallax.

THROW
(1) The distance from the projector to the screen; (2) the positioning and alignment of lights in order to achieve the required lighting effect on the field of action.

TIE-DOWN
See Tee.

TIGHT SHOT
A shot whose subjects are closely framed, such as a "tight two-shot." (See also Close Shot, Shot, Two-Shot)

TIGHT TWO-SHOT
See Close Shot.

TIGHT WIND(ER)
A film rewinder using a roller, spindle, and hub to wind the film onto a core.

TIGHTEN
See Pull Up.

TILT
The vertical movement of a camera on its tripod. Sometimes called "vertical pan." (See also Pan)

TILT ANGLE
(Also Dutch Tilt Angle)
As a general rule the plane of a shot is parallel to the plane of the ground if outdoors, or of the floor if indoors. In a tilt angle shot the camera is tilted out of the plane of the horizontal, whether to right or left. (See also Camera Angle)

TILT SHOT
A shot accomplished while the camera moves up or down on its tripod. (See also Shot)

TIME
To decide which printer light is best for a shot.

TIME BASE SIGNAL
A method of expediting film and sound synchronization for a workprint. A signal is recorded on the edge of film in a camera to correlate with a signal recorded on the magnetic sound track.

TIME LAPSE
See Stop Motion.

TIME-LAPSE CUTAWAY SHOT
A shot interjected between sequential action shots in lieu of a jump cut, to distract the audience momentarily, so that when the main action returns to the screen, time has advanced in the scene without explanation being necessary.

TIME-LAPSE MOTOR
A motor and related equipment attached to a camera, allowing single-frame exposure at selected intervals.

TIME-LAPSE PHOTOGRAPHY
Through the use of single framing at predetermined intervals, events

that take days, weeks, or months to transpire can be compressed into seconds or minutes of screen time. The growth of plants and the opening of flowers are classic examples. (See also Fast Motion, Single-Frame Shooting, Stop Motion)

TIME SLOT

An alloted period of air time for a specific program or purpose; a commercial time slot, in which an advertisement is aired; network time and television (and radio) time at all stations is broken down into time slots.

TIME TRANSITION

A bridge (audio, visual, or both) that indicates the passage of time between two consecutive scenes. (See also Bridge)

TIMED WORKPRINT

A workprint made using various printer lights to correct under- and over-exposure.

TIMER

A person whose responsibility it is to examine each shot in a film and determine how much printer light and color correction are needed.

TIMING

(1) The process of reading a script and tracking the time on a stop watch to determine the length of the overall script and its individual scenes. The decision to lengthen or shorten a script is based on this timing (See also Stop Watch); (2) the use of pauses in a performer's delivery to achieve a dramatic effect; (3) the first steps in printing a film to determine color correction and printer light settings. (See also Grading)

TIMING CARD
(Also Timing Tape)

A punch card or paper tape upon which are marked printer control light changes and color corrections.

TITLE

The name of a film or television show (See also Working Title); (2) the list of credits that appear either at the beginning (See also Opening Credits), or at the end of a film (See also Closing Credits). (See also Intertitles)

TITLE CARD

Cardboard or cel on which a credit or title has been printed.

TITLE MUSIC

Background music used only with the opening and closing credits of a film, as distinguished from the score.

TITLE STAND (Also Titler)

A combination camera support and title card holder that allows easy photography of the cards.

TONE

An unwavering pitch or audio signal of one frequency.

TONED PRINT

A print in which certain scenes have been color-washed to fit the mood, such as a sepia tone for a candlelight or "olde time" effect or a blue tone for a night effect.

TONES AND BAR
The color stripes and sound tones forming the test pattern seen on camera before a show is taped or on television before a broadcast day begins. These test audio and video clarity.

TONGUE
To move the boom of a dolly to the right or left; also, the command "tongue right" or "tongue left."

TOP LIGHTING
Light directed down onto the performers from sources placed above their heads.

TOP-100 MARKET
See Major Market.

TRACED MATTE
A matte made from the tracing of the action found on a single frame of film. (See also Matte)

TRACK
(1) A term given a zoom lens with the ability to keep the same point in focus throughout its entire zoom range; (2) the ability of a camera dolly to retrace a straight line movement (see Trucking). (See also Soundtrack, Zoom Lens)

TRACK LAYING
Positioning and leveling dolly tracks in preparation for a shot.

TRACKING
See Trucking.

TRACKING SHOT
A shot made while the camera and its dolly are moving.

TRADES
Daily and weekly newspapers that specialize in reporting news of the entertainment industry. The daily *Variety* and the *Hollywood Reporter* are two of the most widely read.

TRAILER
Short excerpts or clips from a film shown in theaters to promote its release, also called previews. There are three types of trailers: (1) theatrical—a short preview of a new film a few weeks prior to its release; this trailer is usually 2½ to 3 minutes long; (2) teaser—a brief preview, usually 1½ minutes long, of a film that will be released in the near, not immediate, future; (3) cross plug—a preview of a film currently showing in the theater's chain or complex. (See also Clip, Preview)

TRAINING FILM
An instructional film that teaches a particular skill or imparts specific information. (See also Educational Film, Instructional Film)

TRANSFER SOUND
To rerecord or duplicate sound from one recording to another. For example, to rerecord sound from a 16mm magnetic master to 16mm optical track.

TRANSFER TITLE LETTERS
The use of heated foundry type to transfer embossed title letters from transfer paper to title boards of cels.

TRANSITION
A connecting passage from one distinct scene to another by one of a

variety of methods: including another scene (transitional scene); optical effect; narration; or music. (See also Bridge, Time Transition)

TRANSITION FOCUS
A camera technique in which the picture is taken out of focus and then immediately put back into focus. This may be done by either one or two cameras.

TRANSPARENCY
A positive image applied to a backing made of transparent material, such as acetate or glass. Usually used for projection, such as vugraphs.

TRAPEZE
A long metal frame usually running from one side of the stage to another, which mounts suspended lights firmly and safely.

TRAVELING MATTE
A strip of matte film that travels through the printer along with the print film. The image on the matte film is transferred to the print film creating space, light, or other special images. (See also Matte)

TRAVELING MATTE PRINTING
The process of printing film with the use of a traveling matte.

TRAVELING SHOT
See Moving Shot.

TRAVELOGUE
A documentary-type film in which the culture, scenery, and general experience of an unusual place or foreign country is depicted.

TREATMENT
A more definitive version of a story line. A treatment follows the outline, but precedes the script. (See also Outline, Script, Story Line)

TREBLE ROLL-OFF
Gradual reduction of the intensity of the high frequencies on a sound track.

TREE
See Christmas Tree.

TRENCH(ING)
A trench dug in which tall actors can stand when it is necessary to be shorter, or for props or cameras in order to produce the desired height effect for shooting. (See also Apple Box)

TRIACETATE BASE
A type of film safety base.

TRIACETATE FILM
A type of film with a triacetate base.

TRIAL COMPOSITE PRINT
See First Trial Print.

TRIANGLE
See Tee.

TRIGGER FILM
A film produced to stimulate discussion about its content.

TRIMS (Also Out Takes)
Unused, leftover pieces of film

that were processed but not edited into the final composite print. Unused trims can be stored in a trim bin for future use if the editor changes his mind about how the film was edited.

TRIM BIN
A storage bin for holding the cut pieces of film, trims, or out takes made during editing.

TRIPACK (Also 3-EMULSION FILM)
Film manufactured with three emulsion layers.

TRIPOD
A camera support consisting of three adjustable legs and a means of mounting the camera on top of the stand.

TRIPOD HEAD
A camera mount, fitted to a tripod, that allows the camera to rotate smoothly.

TRIPTYCH
A film or shots in a film in which three separate images are shown side by side simultaneously. Abel Gance's silent film, *Napoleon*, is famous for the use of this technique.

TROMBONE
A tubelike clamp used to hang small studio lights from the tops of set walls.

TRUCKING (Also Tracking)
Moving the camera and dolly either horizontally past the subject or alongside the moving subject while filming.

TRUCKING SHOT
A take accomplished while the camera and dolly are in motion. (See also Moving Shot, Shot)

TTL
Abbreviation for Through-the-lens Reflex.

TUBBY
See Boomy.

TUNGSTEN-HALOGEN LAMP
Another term for "quartz lamp."

TUNGSTEN LAMP
Any incandescent lamp with a tungsten filament.

TUNGSTEN LIGHT
Light emanating from an incandescent lamp using a tungsten filament.

TUNGSTEN RATING
(Also Tungsten Index, Tungsten Speed)
A number used in conjunction with the light meter to indicate a film's sensitivity to artificial light as specified by the manufacturer, laboratory, or director of photography after tests.

TURNAROUND
(1) A term describing a project that has been in development at a production company or network but is dropped and is then picked up and completed by another studio. (See

TURRET

also In Turnaround); (2) the time necessary to break down one set on a soundstage and set up another. (See also Flip-Over)

TURRET
A rotating lens mount on the front of a camera, capable of holding two or more lenses. It may be rapidly rotated into position for filming.

TV
Abbreviation for television.

TV CROPPING
See Television Cutoff.

TV MASK
See Television Mask.

TV PRINT
A specially balanced color release print used for xenon projection.

TVQ
A rating service purported to assist the television industry by assessing the marketability of approximately 1,000 TV stars by over-the-phone, nationwide surveys of approximately 3,000 people.

TWOS
Filming two frames for each change in animation artwork instead of just one frame. (See also Animation Camera)

TWO-BLADED SHUTTER
A shutter on a projector, consisting of two blades and two openings, which eliminates flicker by projecting each frame twice.

TWO-INCH (2″)
Refers to a width of magnetic video (or audio) tape most commonly used to record the master in video (or audio) tape production.

TWO PLATE
(Also Four Plate)
A term describing the capacity of a console editing machine. The more plates a machine has, the more film and soundtrack rolls it can play simultaneously.

TWO-SHOT
A shot in which two people are framed, usually from the waist up. (See also Shot, Tight Shot)

TWO-TRACK STEREO
Sound recorded through a two-microphone pick-up pattern or a multiple-microphone pick-up pattern that, when played back, has a definite left-channel and right-channel sound distribution.

TWO-WALL SET
A set consisting of two walls that, when assembled, form a corner. (See also Set)

TYPE CASTING (Also Typage)
The process of casting someone in a role who looks the part of the character or has the right personality, body type, voice, or other feature that precisely fits the role, as opposed to casting someone in that role for his or her talent or acting ability.

240

UHF

The frequency band from 300 to 3,000 MHz, including television channels 14 through 83. (See also VHF)

ULTRASONIC FILM CLEANER

A device that easily cleans film using high frequency sound waves to remove the dirt. (See also Film-cleaning machine)

ULTRAVIOLET
(Also Black Light)

Invisible light of a shorter wavelength than visible violet; it can be detected by film emulsions unless an ultraviolet filter is used to stop this radiation.

ULTRAVIOLET FILTER
(Also Haze Filter, UV Filter)

A filter that absorbs ultraviolet light that can be used during filming as well as in the developing process to eliminate ultraviolet light exposure on the finished film.

ULTRAVIOLET MATTE

A special-effects matte that is made by filming the actors or subjects in front of a translucent screen, backlit by ultraviolet light; another matte can then be used to replace the translucent screen and thereby create any background for the actors or subjects. (See also Matte, Special Effects)

UMBRELLA

An umbrellalike device made of white or silver cloth used to diffuse light. (See also Diffusers)

UNDER

An editing or sound production directive to reduce the music or sound effects being used to a lower level, at which it will be heard as background and not distract from the main sound or voice being recorded. Often soft music is heard under a sad scene; in which case the main voices or sound has probably been looped over a controlled track that contains the background sounds.

UNDERCRANK
(Also Undercranked)

A term that refers to slowing down the speed of the camera. This creates the effect of speeding up the action. The term is a carry over from silent days when all cameras were operated and controlled with manual hand cranks. (See also Fast Motion)

UNDERDEVELOPMENT

Not allowing the film sufficient time in the developer in which to bring up the image, resulting in a faded, unclear picture.

UNDEREXPOSE

Preventing sufficient light from reaching film exposed in a camera or printer; exposing or printing so that the image is darker than normal.

UNDERGROUND FILM
An experimental, usually low-budget film done by independent, often amateur filmmakers. (See also Experimental Film)

UNDERPLAY
To deliver a line, scene, or entire performance in a low-key manner, sometimes with obvious restraint.

UNDERSHOT
An amount of film insufficient for satisfactory coverage and editing.

UNDERWATER HOUSING
A watertight camera container for underwater photography.

UNDEVELOPED
Film that has not yet been developed.

UNEXPOSED
Raw stock that has not been exposed to light and has not been used in a camera or printer.

UNIDIRECTIONAL MICROPHONE
A mike with greater sensitivity to sound from one direction. (See also Directional Microphone, Microphone Pick-Up Pattern)

UNILATERAL TRACK
A soundtrack capable of generating sound from only one side of a variable area soundtrack.

UNIT
A group of production people assigned to a specific technical area of production; such as the special effects unit or second unit photography.

UNIT MANAGER
The behind-the-scenes expediter, responsible to the production manager, who is responsible for any or all of the following tasks: transportation, housing, payroll, supplies, and meals.

UNIT OF ACTION
A specific incident of action that takes place at a specific place and time.

UNIT SET
A set that can be assembled in various ways, so that different scenes can be played with minimal changes in lighting and set dressing.

UNIVERSAL FOCUS
The sharp, clear focus of all images, whether near or far. (See also Focus)

UNIVERSAL LEADER
Film leader manufactured with specific standardized marks to facilitate threading in projectors and changeovers. (See also Leader, SMPTE Universal Leader)

UNMODULATED TRACK
Any silent portion of a soundtrack. (See also Soundtrack)

UNSQUEEZE
Using an anamorphic lens to normalize the laterally compressed images inherent in wide-screen pho-

tography. (See also Anamorphic Lens)

UP

(1) "Up"—a directive to increase the volume level or intensity on any part of a production, for example, "bring up the lights" or "up the music"; (2) said of an actor who has forgotten his lines.

UP SHOT

A shot taken from a low angle with the camera slanted up. (See also Shot)

UPSIDE-DOWN SLATE
(Also Tail Slate)

A slate and clapstick pop made at the end of a scene instead of at the beginning. The slate is held upside down to indicate it is the end of the scene as opposed to the beginning.

UP STAGE

(1) The back of a stage; (2) action by a performer or element of production that causes the audience's attention to be shifted away from the main action; (3) a directive meaning away from the audience.

V

VARIABLE AREA TRACK
An optical soundtrack that consists of a narrow transparent area on a strip of film, proportionally varied in width to the modulating signal. (See also Modulation, Optical Soundtrack, Soundtrack)

VARIABLE DENSITY SOUNDTRACK
(Also Variable Density Track)
An optical soundtrack on which signals are represented by variations in density and recorded on the edge of the film. The track is produced by varying the amount of light that strikes the film as the film moves through the projector. (See also Optical Soundtrack, Soundtrack)

VARIABLE FOCUS LENS
See Zoom Lens.

VARIABLE HUE SOUND RECORDING
Sound recording that uses a photographic audio track with variations in color as opposed to variations in monochrome density, still in the experimental stages.

VARIABLE OPENING SHUTTER
(Also Variable Shutter)
A motion picture camera shutter whose opening can be changed either before or during operation, thus allowing for fades to be made. (See also Shutter)

VARIABLE SHUTTER CONTROL
A knob that controls desired changes in the opening of a variable shutter on a motion picture camera. (See also Variable Opening Shutter)

VARIABLE SPEED CONTROL
Any device connected to a camera or its motor to vary the speed.

VARIABLE SPEED MOTOR
(Also Wild Motor)
A motor capable of operating at various speeds.

VAULT (Also Film Vault)
A secure place where films are stored, especially originals, in which the humidity and temperature are carefully controlled.

VELVETING SOUND
(Also Gloving Sound)
A technique of cleaning a soundtrack by wiping it between two pieces of soft cloth.

VERTICAL PAN (Also Tilt)
A vertical camera movement in which the camera is pivoted up and down. (See also Pan)

VERTICAL WIPE
See Wipe.

VHF
The television frequency band

that includes the most commonly used channels. It is separated into the low VHF band, channels 2-6, and the high band, channels 7-13. (See also UHF)

VHS
A half-inch videotape recording and playback system. One of two systems available to the public, the other being Betamax (See also Betamax). VHS has a larger cassette than Betamax as well as a difference in how the tape is fed into the videotape playback machine.

VIDEO
(1) The picture portion of a program; (2) the visual element of any form of visual communication; (3) a colloquial abbreviation for videotape; (4) a visual demo; (5) a complete visual production, for example, a rock video.

VIDEO DISK
A recording and playback disk on which video and audio are stored.

VIDEO RECORDER
A recording and playback unit that records television picture and sound. The most common sizes use 1/2-inch videotape (see also VHS and Beta), 3/4-inch, 1-inch, and 2-inch widths.

VIDEO UP
An editing term, indicating that the luminescence in the picture should be brightened, or brought up.

VIDEOCASSETTE
A cartridge containing a videotape loop, available in various lengths and sizes. The most common cassette sizes are 1/2-inch, 3/4-inch, 1-inch, and 2-inch.

VIDEOGRAPHICS
Television images as created electronically by the console or a computer. (See also Graphics)

VIDEOTAPE
Flexible film coated with iron oxide upon which sound and picture are recorded. Available in sizes of 1/2-inch, 3/4-inch, 1-inch, and 2-inch, on reel or cartridge.

VIEWER(S)
(1) An optical, mechanical device that allows examination of a motion picture film enlargement; (2) the audience.

VIEWFINDER
(Also Independent Viewfinder, Monitor Viewfinder)
The part of a camera that optically shows the exact area being photographed; sometimes built into the camera, sometimes a separate unit. External viewfinders must be corrected for parallax to be accurate. (See also Director's Finder, Ground-glass Finder, Optical Viewfinder, Reflex Viewfinder, Zoom Finder)

VIEWFINDER MISMATCH
An incorrect viewfinder objective lens or mask for the camera lens being used.

VIEWFINDER OBJECTIVE LENS
The image-forming lens at the front of an optical monitoring viewfinder. (See also Lens)

VIEWING FILTER
(Also Contrast Glass)
A filter through which a camera operator evaluates the lighting. It shows the contrast and shadow detail that will appear on various color and black-and-white films. (See also Filter)

VIGNETTING
A technique in which a subject is framed by blurring the area around it, whether created intentionally or accidentally by a poorly placed flag or sunshade; to reduce the sides of a picture while the image at the center is undisturbed; also, any picture so framed.

VISCOUS PROCESSING
A method of film processing in which slow-flowing chemical baths are used.

VISIBLE SPECTRUM
Color that can be seen with the naked eye; the bands of colored light between infrared and ultraviolet that can be seen when white light is refracted, i.e. by a prism.

VISUAL DEVICES
The various methods used in obtaining special visuals, for example, the use of miniatures, models, film clips, or special video effects. (See also Special Effects)

VISUAL EFFECTS
Special effects created with visual images. (See also Special Effects)

VISUAL PRIMARY
A term used to express that the image in a shot is to be more dominant than the sound.

VISUALS
(1) Motion picture images; (2) an action or gesture.

V.O., V/O, VO
Abbreviations for Voice Over.

VOICE CUE
A verbal cue usually given by the director or an assistant for an actor to begin an action or make an entrance. The cue can also be a word of dialogue in the script. (See also Cue)

VOICE OVER (VO, V.O., V/O)
(1) In video, when a voice is heard but the speaker is not seen; (2) in audio, when one speaks over a music or sound effects track; (3) any off-screen voice. (See also Interior Monologue, Narration)

VOICE TEST
A test recording to see if a narrator's or actor's voice will record satisfactorily.

VOLUME
The level of loudness or softness in audio recording, reproduction, or playback; sound amplitude.

VOLUME CONTROL
See Gain Control.

VTR
Abbreviation for Videotape Recorder.

VU METER

A meter with a needle that moves on a graph in accordance with the sound input. An integral part of modern sound reproduction, the VU meter indicates the recording level and variations in volume; it is found on almost all tape recorders and playback units.

VU GRAPHS

See Transparency.

W

WALK-THROUGH
An early stage of rehearsal that takes place on the set. Usually no dialogue is read, but the performers and director work through the logistics of the action to be shot. Also, when the director works the scene out in detail with the crew, stand-ins sometimes are used to replace the actors during the more tedious and difficult shots. (See also Run-Through)

WALKIE-TALKIE
A limited range portable radio transmitter-receiver.

WALL BRACKET
A wall support for a lighting instrument.

WALL PLATE
A small, pedestal light support.

WALL POWER
Standard current from a wall socket, 60-cycle, 120-volt alternating current.

WALL SLED
A light support designed for use on a set wall, or flat, and consisting of a simple, metal bracket.

WARDROBE
(1) All clothing, costumes, and related accessories worn by performers in a production; (2) the department on a studio lot or area in production specifically designated to handle and store all articles or clothing and accessories used in production.

WARDROBE TRUCK
A vehicle designated to transport clothing, costumes, and their accessories from the wardrobe department or place of storage to the set or film location; the truck also contains equipment needed to repair costumes.

WARDROBE WOMAN (MAN)
The person who delivers the proper clothing, costume, and accessories to the performers; is responsible for their upkeep; and assists the performers in dressing, if necessary.

WARM IMAGE
Dominant red/yellow tones created purposely with a filter, lights, a special lens that is capable of absorbing the blue end of the spectrum, or the addition of red/yellow in the printing phase of development. (See also Cool Image)

WARM-UP
The relaxing and loosening-up process created to make a live audience more responsive to the show they are about to see. The warm-up is usually a ten- to thirty-minute ad-

libbed performance by a specially hired announcer (see also Warm-up Announcer) and takes place just prior to the beginning of a live show. During a warm-up, an audience is told what to expect and what is expected of them, for example, applause at certain points, or no whistling as the sound system is too sensitive.

WARM-UP ANNOUNCER

A specialized field of performance in which the person's sole duty is to warm up an audience before they view a show. (See also Warm-Up) In addition to telling jokes, the warm-up announcer usually answers questions from the audience about the production and its performers, involves the audience in what they are about to see, and imparts information necessary to their enjoyment and participation.

WASH

The final rinse in a water bath of developed and fixed film.

WATERTIGHT CAMERA

See Underwater Housing.

WATERSPOTS

A visual defect on film that appears as blotches, droplets, or stains caused by improper drying of the film in the lab.

WAVE MACHINE
(Also Wave Generator)

An electrically powered special effects device used to create waves of water for filming. It is a wedge-shaped box that is mechanically forced up and down in a large tank of water; height and rhythm of the waves are adjustable.

WAXING

(1) The application of wax to the edges of a release print in order to smooth its course through the projector, thereby preventing damage to the film's emulsion; (2) the wax solution or wax product applied in waxing.

WEAVE

The undesirable sideways motion of film in the camera or projector gate.

WEB

The total components that as a whole comprise a broadcasting unit, including business offices, affiliates, studios, and all points of operation; a broadcast, syndication or cable network, a group of stations that have a connection with the program originators.

WEDDED PRINT

See Composite Print.

WEENIE

Any gimmick, hook, or twist used in a plot to give it an unpredictable turn. (See also Plot Gimmick)

WEST

The term used to designate the left side of an animation field chart. (See also Animation)

WESTERN

A film set in the West, generally during early frontier days. A modern Western takes the Western theme and transposes it into current times. The main key of a true Western lies in the elements of plot in which a hero lives by a code of ethics, based on the need to survive in a frontier environment, and in the course of the action faces many foes. *Star Wars* has been referred to as a modern Western. (See also Horse Opera, Oater)

WET GATE PRINTING
(Also Liquid Gate Printing)

A process in the printing of film in which the footage passes through pads filled with fluid. The fluid fills any scratches on the film, thus preventing them from appearing in the final print.

WGA

Abbreviation for Writers Guild of America.

WHEELED TIE-DOWN
(Also Wheeled Triangle, Wheeled Tee)

A mobile tee allowing movement of the camera while keeping it stabilized on its tripod; in some cases, making dolly shots possible. (See also Tee)

WHIP PAN

A rapid pan of the camera from one subject to another. (See also Pan)

WIDE-ANGLE DISTORTION
(Also Wide-Angle Effect)

An exaggerated image effect created by photographing objects through a wide-angle lens at close range, resulting in extreme foreshortening and roundness of the image, as well as an unnatural lengthening of images near the edge of the frame.

WIDE-ANGLE LENS
(Also Short Focal Length Lens)

Any lens with shorter-than-normal focal length in any given frame size, with a wider-than-normal field of view.

WIDE-ANGLE SHOT

A shot made with a wide-angle lens in which more of the action field is seen than would be seen in a shot made with a normal lens. A wide-angle camera lens sees more closely what the normal eye would see, whereas most lenses cut the peripheral area of a scene.

WIDE-SCREEN

Any screen wider than the normal aspect ratio of 1 to 1.3 of early screens. The standard wide-screen aspect ratio used most often in today's theaters is 1 to 1.65.

WIDE-SCREEN RELEASE PRINT

A wide-screen print, cropped, whose frame area has an aspect ratio of 1 to 1.65.

WILD LINES

Unsynchronized lines of dialogue added after the film has been shot, during the editing and sound mixing process.

WILD CAMERA

A camera that is not connected with the sound recorder and thus shoots silently.

WILD MOTOR
(Also Variable Speed Motor)

A camera motor with adjustable speeds that does not run in exact synchronization.

WILD PICTURE
(Also Wild Shot)

Footage that is filmed without synchronous sound.

WILD SOUND
(Also Wild Recording)

Recording of sounds not in sync with the picture, such as various sound effects and background noise.

WILD TRACK

An audio track made from sounds that are recorded independent of the film. Sometimes called a "floating track."

WILD WALL

A wall that is capable of being removed from the set. (See also Cheat)

WIND (UP)

(1) The emulsion position on a single-sprocket film roll (see also A-wind, B-wind); (2) the process of spooling or winding film on a reel or core; (3) to turn a winding key or crank on a camera, coiling the spring mechanism, which in turn allows the camera to operate; (4) to finish shooting, rehearsing, or any process in active production.

WIND GAG

A screen, often resembling a foam rubber sock, that encases a microphone, muffling the sound made by the wind when shooting out of doors.

WIND MACHINE

A large fan that creates special effect winds ranging from mild gusts to simulated hurricane-force gales.

WIPE

A special effects transition from one scene to another in which one scene is slowly "wiped" off the screen as the image of the next scene progressively appears. The two scenes are separated by a moving dividing line that wipes away one scene as would a windshield wiper. The most common wipe is vertical, with the scene-dividing wipe line moving from left to right; other wipes are diagonal, iris, spiral, and "atomic bomb." "Wipe to black" means that a dark screen follows the wipe line so that no picture is left when the wipe is complete. (See also Push Off, Soft Edge Wipe)

WIRELESS MICROPHONE
(Also Radio Microphone)

A wireless mobile microphone capable of recording sound through a receiver, using a short-distance transmitter, thereby eliminating the necessity of a microphone cable. (See also Microphone)

WORKPRINT

The first print made from the original film footage. This will be used in the editing process to determine what cuts will be made and where music and dubbing will be added. The use of a workprint upon which to do these trial cuttings will protect the original print from damage and, when the workprint has been fine cut, it will act as a guide for the editor in conforming the original. (See also Assemble, Rough Cut)

WORK LIGHT
A studio light that provides illumination for any activity outside of actual filming.

WORKING DISTANCE
The ideal distance between microphone and performer to assure optimum pick-up of sound.

WORKING DRAWINGS
Scale drawings and diagrams of flats indicating dimensions and providing instructions for construction of the set. (See also Flat)

WORKING LEADER
Blank film at the beginning of a roll used for threading through equipment during the editing process; and upon which sync marks are written. (See also Leader)

WORKING TITLE
A tentative title assigned to a project during production, until a final title is chosen. (See also Title)

WOW AND FLUTTER
(Also Flutter Sound)
Distortions occurring in sound during recording or playback, generally caused by variations of the speed at which the tape is moving.

WRANGLER
(Also Livestock Man)
The crew member who is responsible for the care of all livestock used in a production, whether they be dogs, bees, fish, or wild animals.

WRAP
The finish of principal photography of a production.

WRAP PARTY
A celebration of the completing of the taping season (television series) or of the principal photography of a film, which usually includes the cast, crew, and all those involved in the production.

WRITER
The person who creates and writes, rewrites, or polishes the story, treatment, or script for production.

WRITERS GUILD OF AMERICA (WGA)
One of the major unions in the industry, founded in 1954 to represent writers primarily for the purpose of collective bargaining in its jurisdiction of screen, television, and radio. The Guild has two branches: Writers Guild of America, West, Inc. for writers west of the Mississippi, and Writers Guild of America, East, Inc. for those east of the Mississippi. Basically, the WGA establishes minimum wages paid to writers, the credits or billing a writer is to receive, and arbitrates differences among writers and producers.

WRONG-READING
Any reversed photographic image that appears backward (mirror image) on the screen. (See also Right-reading)

X

X's
A script abbreviation indicating that the actor is to move across the set.

X LIGHTING (Also Double Key)
A lighting technique used on two people speaking to each other. The key light for each acts as the backlight for the other. (See also Back Light, Key Light)

XCU (Also ECU)
Abbreviation for term extreme close-up.

XENON LAMP
A high intensity lamp used in projection equipment, that is filled with xenon gas.

XLS
Abbreviation for the term extreme long shot.

YOKE

A moveable U-shaped part of a spotlight, located between the hood and the swivel, which allows the light beam to move on a vertical path.

Z

Z CORE
A plastic core, 3 inches in diameter, used to hold rolls of 16mm film longer than 400 feet.

ZB
Abbreviation for Zoom Back.

ZELDA
The head and shoulders of a mannequin, placed on the stage in front of the cameras and used by the camera operators to focus.

ZEPPELIN WINDSCREEN
A perforated and elongated tube which eliminates wind noise when slipped over a shotgun microphone.

ZERO CUT CONFORMING AND PRINTING
Preparing A and B rolls for printing by overlapping original shots by several frames. The change from one roll to another to match the edited workprint is done automatically by the printer. Image definition at the cut is better in the print because a splice is not needed as with overlapping, and a splice-mark is therefore eliminated.

ZERO FRAME
A reference frame, the edge of which is level with a key number. This frame, along with those occurring before the next key number appears, is used to accurately identify any individual frame, which is essential during lab and optical effects work.

ZI
Abbreviation for Zoom In.

ZIP PAN
See Swish Pan.

ZOOM
The act of changing the focal length, or distance, being filmed, enlarging or reducing, bringing the action closer or further from the viewer.

ZOOM BLIMP
A covering, designed especially for a zoom lens, that reduces the camera noise when fitted over the lens. (See also Blimp)

ZOOM DRIVE
An electrically powered zoom lens, controlled by the camera operator to zoom in or out rapidly or slowly, making zooms smoother than those made by hand.

ZOOM FINDER
Mostly replaced by the newer reflex viewfinders, this older piece of equipment shows the variable field covered by a zoom lens in a separate, parallax viewfinder. (See also Reflex Viewfinder)

ZOOM IN (Also ZI)

Increasing the focal length of the zoom lens during a shot, resulting in the gradual magnification of the images when projected on a screen.

ZOOM LENS
(Also Variable Focus Lens)

A variable focal length lens capable of changing its focal length, thereby changing the magnification of the subject. Unlike zoom lenses for still cameras, zoom lenses for motion-picture cameras are able to track the subject and effectively go from a long shot to a tight close-up in a smooth, clear, continuous flow without changing the focus of the lens. to the untrained eye, it appears as if the camera is moving; however, only the lens is. (See also Joy-Stick Zoom Control, Macrozoom Lens, Track)

ZOOM OUT
(ZO, Zoom Back, ZB)

Decreasing the focal length of the zoom lens during a shot, resulting in the subject becoming smaller when projected on a screen.

ZOOM RANGE

The range of a zoom lens from its longest focal length to its shortest.

ZOOM SHOT

A shot made with a varifocal lens whose focal length is changed during the shot, i.e., images become progressively larger or smaller, closer or further away. (See also Shot)

REF PN 1992 .18 .E57